Practical Machine Learning for Streaming Data with Python

Design, Develop, and Validate Online Learning Models

Sayan Putatunda

APress®

Practical Machine Learning for Streaming Data with Python: Design, Develop, and Validate Online Learning Models

Sayan Putatunda
Bangalore, India

ISBN-13 (pbk): 978-1-4842-6866-7   ISBN-13 (electronic): 978-1-4842-6867-4
https://doi.org/10.1007/978-1-4842-6867-4

Managing Director, Apress Media LLC: Welmoed Spahr
Acquisitions Editor: Celestin Suresh John
Development Editor: James Markham
Coordinating Editor: Shrikant Vishwakarma

Cover designed by eStudioCalamar

Cover image designed by Pexels

Distributed to the book trade worldwide by Springer Science+Business Media LLC, 1 New York Plaza, Suite 4600, New York, NY 10004. Phone 1-800-SPRINGER, fax (201) 348-4505, e-mail orders-ny@springer-sbm.com, or visit www.springeronline.com. Apress Media, LLC is a California LLC and the sole member (owner) is Springer Science + Business Media Finance Inc (SSBM Finance Inc). SSBM Finance Inc is a **Delaware** corporation.

For information on translations, please e-mail booktranslations@springernature.com; for reprint, paperback, or audio rights, please e-mail bookpermissions@springernature.com, or visit http://www.apress.com/rights-permissions.

Apress titles may be purchased in bulk for academic, corporate, or promotional use. eBook versions and licenses are also available for most titles. For more information, reference our Print and eBook Bulk Sales web page at http://www.apress.com/bulk-sales.

Any source code or other supplementary material referenced by the author in this book is available to readers on GitHub via the book's product page, located at www.apress.com/978-1-4842-6866-7. For more detailed information, please visit http://www.apress.com/source-code.

Printed on acid-free paper

To my family
—Dr. Sayan Putatunda

Table of Contents

About the Author

Dr. Sayan Putatunda is an experienced data scientist and researcher. He holds a PhD in applied statistics and machine learning from the Indian Institute of Management, Ahmedabad (IIMA), where his research was on streaming data and its applications in the transportation industry. He has a rich experience of working in both senior individual contributor and managerial roles in the data science industry with companies such as Amazon, VMware, Mu Sigma, and more. His research interests are in streaming data, deep learning, machine learning, spatial point processes, and directional statistics. As a researcher, his work has been published in top international peer-reviewed journals. He has presented his work at various reputed international machine learning and statistics conferences. He is a member of IEEE.

About the Technical Reviewer

Manohar Swamynathan is a data science practitioner and an avid programmer, with more than 13 years of experience in various data science-related areas, including data warehousing, business intelligence (BI), analytical tool development, ad hoc analysis, predictive modeling, data science product development, consulting, formulating strategies, and executing analytics programs. Manohar's career coves the life cycle of data across different domains, such as US mortgage banking, retail/e-commerce, insurance, and industrial IoT. He has a bachelor's degree with a specialization in physics, mathematics, and computers. His master's degree is in project management. He's currently living in Bengaluru, the silicon valley of India.

Manohar is the author of Mastering *Machine Learning with Python in Six Steps* (Apress, 2019). You can learn more about his various other activities on his website at www.mswamynathan.com.

Acknowledgements

I would like to thank all those who have contributed to bringing this book to publication for their help, support, and input. I would like to give a heartfelt dedication to my mother (Mrs. Mili Putatunda) and my father (Mr. Supriya Putatunda) for their lifelong love, support, and encouragement in all aspects of my life. My special thanks to my wife (Mrs. Srijani Putatunda) for her support and encouragement. I am indebted to my thesis adviser, Prof. Arnab Kumar Laha, for introducing me to this wonderful research topic (i.e., analysis of streaming data) during my PhD days at IIM Ahmedabad and for having countless captivating and patient conversations with me.

I would also like to thank and appreciate the efforts of the editors of this book, Celestin John, James Markham, and Shrikant Vishwakarma, at Apress/Springer for their essential encouragement.

Dr. Sayan Putatunda

Kolkata, India

January 2021

Introduction

Advancements in technology have made the continuous collection of data possible. We are inundated with data in various daily transactions (e.g., POS transactions in retail outlets like Walmart, Target, etc.), sensor data, web data, social media data, stock prices, search queries, clickstream data, and so forth. This is all a source for high-velocity data; that is, streaming data.

This book is a quick-start guide for beginners to learn, understand, and implement various machine learning models on streaming data to generate real-time insights using Python. Streaming data is defined and infinite and continuous inflow of data at a high pace. Most of the machine learning models that we deal with are in the batch context. However, batch learning models are not suitable for handling streaming data as multiple passes over the data are not possible. The batch models may soon become outdated due to *concept drift* (i.e., the data distribution changes over time). Other challenges of streaming data include memory limitations and the high speed of data accumulation, which led to the development of a different class of methods, known as *incremental* or *online learning methods*.

The industry's adoption of various online learning/incremental learning methods has been quite slow over the years, but things are now changing quickly due to the vast applications of real-time machine learning. So, these techniques should be a part of a data scientist's repertoire!

This book is suitable for data scientists, machine learning engineers, researchers, software engineers, academicians, and data science aspirants with basic programming skills in Python and keen on exploring machine learning with streaming data for a career move or an enterprise/academic project.

INTRODUCTION

This book focuses on an end-to-end accelerated track delving into a holistic approach to develop concept drift detection algorithms, supervised learning algorithms (for regression and classification tasks), and unsupervised learning algorithms implemented in Python. Overall, this book comprises four chapters.

Chapter 1 introduces the concept of streaming data, its various challenges, some of its real-world business applications, and various windowing techniques. This chapter also introduces the concepts of incremental and online learning algorithms and the scikit-multiflow framework in Python.

Chapter 2 covers the various change detection/concept drift detection algorithms and their implementation on various datasets using scikit-multiflow.

Chapter 3 discusses the various regression and classification algorithms for streaming data (including ensemble learning) and its implementation on various datasets.

Finally, Chapter 4 introduces unsupervised learning for streaming data. It also provides a brief overview of other open source tools for handling streaming data, such as Spark Streaming, Massive Online Analysis (MOA), and more.

CHAPTER 1

An Introduction to Streaming Data

This chapter introduces you to streaming data, its various challenges, some of its real-world business applications, various windowing techniques, and the concepts of incremental and online learning algorithms. The chapter also introduces the scikit-multiflow framework in Python and some streaming data generators.

Please note that there are some other frameworks available in Python and for machine learning with streaming data. The scikit-multiflow framework is used in this book because I strongly believe that this package (given its wide range of implemented techniques and great documentation) is a good starting point for Python users to pick up online/incremental machine learning techniques for streaming data.

Streaming Data

The world has witnessed a data deluge in recent years. There has been a huge increase in the volume of data generated from various sources. The major data sources are the Internet, log data, sensor data, emails, RFID, POS transaction data, and so forth. The data gathered from these sources can be categorized into structured, semi-structured, and unstructured data. Recent technological advances are a major reason for this "data explosion," making

© Sayan Putatunda 2021
S. Putatunda, *Practical Machine Learning for Streaming Data with Python*,
https://doi.org/10.1007/978-1-4842-6867-4_1

data storage cheaper and a continuous collection of data possible. Almost all companies in sectors such as retail, social media, and IT are facing a data explosion. They are trying to figure out ways to process and analyze the massive data that they are generating and gain actionable insights.

BIG DATA

Most of us have heard of the term *big data*. The McKinsey Global Institute define big data as "datasets whose size is beyond the ability of typical database software tools to capture, store, manage, and analyze" [1]. Big data's basic characteristics can be defined with four words that begin with the letter V: volume, velocity, variety, and veracity. The "big" in big data is not only about the size or volume of the data—the other "V"s also need to be considered.

Advancements in technology have made the continuous collection of data possible. We are inundated with data in various daily transactions (e.g., POS transactions in retail outlets like Walmart, Target, etc.), sensor data, web data, social media data, stock prices, search queries, clickstream data, and so forth. This is all a source for high-velocity data; that is, streaming data. Other sources of continuous data streams include operational monitoring, online advertising, mobile data, and the Internet of Things (IoT).

Streaming data, or data streams, are an infinite and continuous flow of data from a source that arrives at a very high speed. Thus, streaming data is a subset of big data that addresses the *velocity* aspect of big data. Some of the differentiating characteristics between streaming data and static data are the loosely structured, always on, and always flowing of data [2]. Unlike static data, structure is not rigidly defined in streaming data. It is "always on;" that is, data is always available because new data is continuously generated.

As an example of how streaming data is harnessed in business, let's look at online advertising. Social media giants like Facebook collect user behavior data in real time. Facebook has more than two billion active users every month [4], which gives you an idea of the scale of the data being collected.

Facebook earns most of its revenue by showing ads to its users. Many advertisers are onboarded in the platform. Facebook uses user behavior data to select ads that are relevant to its users (i.e., ads that have the maximum chance of interaction in the form of clicks or purchase). They perform other important steps in online advertising (such as placing a "bid" in an "auction"). All of this happens in near real time (i.e., whenever you load your News Feed page in Facebook or interact with other apps, such as Messenger or Instagram).

The Need to Process and Analyze Streaming Data

Streaming data offers real time or near real-time analytics. Since it's ever-producing and never-ending, storing this enormous data and then running analytics on it (as done in batch processing) is not feasible. Streaming data must be analyzed "on the fly" to gain insights in real time or near real time.

Figure 1-1 shows the streaming data analytics process flow. High-velocity data from various sources (sensor data, stock-tick data, etc.) are ingested, processed, and analyzed by a streaming data analytics engine in real time or near real time. The output generated is displayed via dashboards, apps, or any other means.

Figure 1-1. *Streaming data analytics process flow*

Streaming data analytics and insights empower an organization's decision-makers to make value-added decisions in real time. That is why many researchers in academia and industry are working on new methods for real-time analytics. Conventional machine learning algorithms do not work because streaming data provides a new set of challenges.

The next section discusses the challenges of handling streaming data.

The Challenges of Streaming Data

The following are some of the major challenges of handling streaming data.

- Multiple passes on data are no longer possible. Data needs to be processed in one pass.

- Data streams may evolve over time, which is a behavior called *concept drift*. Mining algorithms for streaming data need to address this issue so that the model does quickly become outdated.

- The high speed needed to stream data.

- Higher memory requirements.

New methodologies are needed for the machine learning and data mining algorithms used with streaming data. The batch processing algorithms for static data don't work. Let's look at a simple linear regression method. In batch processing, entire training datasets are available. Linear regression is applied to the whole dataset using OLS (ordinary least squares) to get beta coefficients to find the relation between the dependent and the independent variable (i.e., beta coefficients signify the amount of change in the dependent variable (y) on an average for a unit change in the independent variable (x)).

In streaming data, however, this is not feasible. You never have access to all the data because it is continuously flowing in. Here you need online learning or incremental learning algorithms, which take each example or tuple of data as it comes and updates the parameters (in this case, the beta coefficients). One such implementation is linear regression with SGD (stochastic gradient descent). A deeper look at online learning/ incremental learning techniques is provided in the upcoming chapters.

Importantly, unlike typical big data analysis, streaming data doesn't require more storage because older data is discarded after a while. Smarter and faster algorithms are needed to process and analyze the incoming fast-streaming data. Algorithms should be capable of generating output anytime it's queried.

You should now understand why machine learning with streaming data is different from that of batch data. The next section discusses some of the real-world applications of streaming data.

Applications of Streaming Data

Real-time analytics using streaming data is going to be a game-changer in this era of big data. The mantra for real-time analytics is to get good solutions quickly. The following are some applications or use-cases for streaming data real-time analytics .

- **Fraud detection in real time or near real time**.
  Let's look at Amazon as an example. There are approximately one billion daily transactions on Amazon.com, but not all the transactions are genuine.

- **Real-time recommendations** can lead to impulsive customer purchases, which is a very important proposition for e-commerce firms.

- **Real-time weather detection** is useful for disaster and warning systems.

- **Network management in real time** involves monitoring and configuring network hardware and software to ensure smooth operation. Use-cases include real-time fault detection, monitoring link bandwidth usage in real time, improving network resource utilization, and more.

- **Real-time operations management** include supply chain decision-making, such as when to refuel or reorder.

- **Algorithmic trading using stock tick data.** Monitoring stock trends in real time helps trading-related decision-making.

- **Real-time sports analytics.** Relevant use-cases include player-movement tracking and performance analytics using streaming data sources, such as video feeds and sensor data.

- **Precision agriculture.** A simple use-case is collecting sensor-based data on farm equipment health and minimizing its downtime by predicting when and where equipment failure occurs and making alternative arrangements.

- **Security event monitoring** includes real-time monitoring of log data, host systems, and security devices to flag incidents (e.g., send alerts) such as failed logins, malware activity, and other suspicious activities.

- **Real-time trends** are a feature in Twitter and Google Trends.

- **Automated annotation**. Real-time annotation of objects in a live video feed is a relevant use-case. The same is applicable for a stream of images or speech data.

In the next section, you learn about the various Windowing techniques for analyzing streaming data.

Windowing Techniques

A window can be defined as a snapshot of data—either observation-count-based or time-based [3]. This is a very useful technique in a streaming data context since at no point in time do you have the "entire data" available. A window can effectively handle *concept drift* because more importance is given to recent data points; the older data points are periodically discarded. A batch learning method can be converted to an incremental method by using a *windowing technique* [5]. The types of windowing techniques include the sliding window model, the damped window model, and the landmark window model [6]. The following explains each of these windowing techniques.

- *Landmark window model.* Older data points are not discarded. All data points are accumulated in the window. Figure 1-2 depicts this model. The rectangle represents the window, and the blue circles are the data.

- *Sliding window model.* This is a popular way to discard older data points and consider only the recent data points for analysis. Figure 1-3 depicts this model.

- *Damped window model.* The data points are weighted. Higher weight is given to recent data points. An *exponential fading strategy* is used to discard old data. An aging function $g(t)$ is used.

$$g(t) = 1/2^{\lambda(t_x - t_o)}$$

λ is the fading factor, t_x is the current timestamp, and t_o
is the origin timestamp (i.e., the time that the window
was created). You can influence discarding older
data points by changing the value of λ [7]. A higher λ
value means less weight is given to older data points.
Figure 1-4 shows the exponential fading of data points
in a damped window model over time.

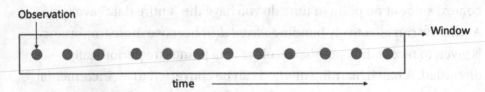

Figure 1-2. *Landmark window model*

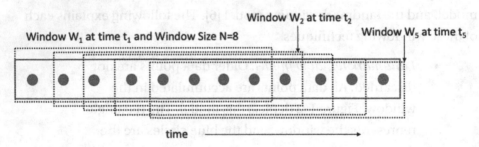

Figure 1-3. *Sliding window model*

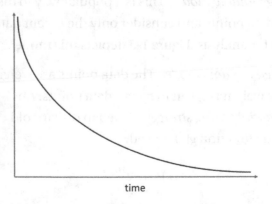

Figure 1-4. *Exponential fading of data points in a damped window model*

Figure 1-2 shows that a landmark window accumulates data points with over time. The window size increases as more data points arrive. This windowing technique requires large memory resources, especially when dealing with large data streams.

In a sliding window model, as shown in Figure 1-3, "time" is the x axis, and the observations (i.e., the blue dots) arrive sequentially from left to right. The leftmost observation is the oldest data, and the most recent observations are on the right.

Figure 1-3 shows that the older points are continuously discarded; only the recent points are considered in the window. Figure 1-3 also shows a window of size = 8 (i.e., it contains eight data points in an instant of time). The window moves in step = 1 (i.e., the window slides in step 1, where the oldest data is discarded).

Also in Figure 1-3, window W_1 at time t_1 contains eight data points. When a new data point arrives at time t_2, the oldest data point in W1 is discarded to form window W_2 with eight data points, including the most recent one. The sliding windows model only considers recent data points; it discards the older ones.

A sliding window is implemented in Python 3, as shown in Listing 1-1. It creates a window of size = 4 and shifts in steps of 1. The input array (y) consists of ten integers from 0 to 9. The sliding window implementation output is shown in Listing 1-2. The sliding windows at the end have None, which indicates an empty value, as no new data point is arriving, but the window size is fixed to 4.

Listing 1-1. Sliding Window Implementation

```
##################################################################
# Import the relevant libraries
import itertools
from itertools import tee
from itertools import zip_longest as zip
```

```
def window(iterations, size):
    n = tee(iterations, size)
    for i in range(1, size):
        for each in n[i:]:
            next(each, None)
    return zip(*n)

y=range(10)

for each in window(y, 4):
    print(list(each))
```
##

Listing 1-2. Sliding Window Function Output

##
```
[0, 1, 2, 3]
[1, 2, 3, 4]
[2, 3, 4, 5]
[3, 4, 5, 6]
[4, 5, 6, 7]
[5, 6, 7, 8]
[6, 7, 8, 9]
[7, 8, 9, None]
[8, 9, None, None]
[9, None, None, None]
```
##

Unlike the sliding windowing technique, the damped window model gives recent data points a higher weight. The older data points are discarded using an exponential fading strategy, as shown in Figure 1-4. The relatively older data points are discarded faster than the newer data points. This way, the window stores both older and newer data points. However, the sliding window technique is much more popular in a streaming data context. It is the primary windowing technique in this book.

In the next section, you learn about incremental learning and online learning algorithms.

Incremental Learning and Online Learning

Most standard machine learning techniques (supervised or unsupervised) are *batch learning* methods. That is, they typically work on static/batch datasets (i.e., non-streaming datasets), where the entire data is available, and multiple passes for any algorithm are possible. However, these batch learning methods become outdated over time when applied to streaming data due to concept drift. There is a need for a different class of machine learning algorithms in the streaming data context that can address its challenges. This class of algorithms is known as *incremental* or *online learning algorithms*.

Incremental learning algorithms work with limited resources and frequently update model parameters whenever a new batch/mini-batch of data arrives. Incremental learning algorithms can handle concept drift inherent in streaming data. Online learning algorithms differ from incremental learning algorithms in that the model parameters are updated whenever a new observation arrives. In online learning algorithms, you don't wait for a mini-batch of the data to arrive to update the model parameters.

An incremental learning algorithm can be approximated by using a batch learner with a sliding window [5]. In this case, the model is retrained every time a new window (comprising data points) arrives. You can apply any batch learner, such as linear regression or logistic regression algorithms, with a sliding window on a data stream to approximate an incremental learning algorithm.

Let's look at a mini-batch gradient descent method and the *stochastic gradient descent* (SGD) method. The gradient descent method is a very popular technique for finding a function's minimum, which in this context is the cost function. The following explains how to do a gradient descent algorithm.

1. Initialize the parameters with random values.

2. Compute the gradient of the objective function.

3. Update the parameters using the gradients.

4. The step size calculation for each feature: step size= gradient * learning rate. The *learning rate* is a hyperparameter that directly impacts the convergence of the algorithm. A very small learning rate takes a very long time for the gradient descent algorithm to converge. A very high value of the learning rate makes the algorithm take large steps down the slope, such that it may miss the global minima.

5. Calculate the new parameters: new parameters = old parameters – step size.

6. Repeat steps 2–5 until the gradient is close to zero.

The gradient descent method is a batch learning method. The entire training data is considered before taking a step in the direction of the gradient. In large datasets, a single update of parameters takes a long time. However, a mini-batch gradient descent algorithm is an incremental learning algorithm where the parameters are changed each time a mini-batch is processed. But SGD algorithm always updates the model parameters for each training observation. The SGD technique is an example of an online learning algorithm.

Let's now look at the linear regression method, which is a linear model to understand the relationship between the target variable (Y) and the predictor variables (X) by computing the beta coefficients for each predictor variables using the *ordinary least square* (OLS) method.

You can implement the linear regression method by using the gradient descent optimization technique. Linear regression is a batch learning technique in which the gradient descent algorithm is applied on the entire training to update the model parameters. However, you can convert this method into an incremental learner by using the mini-batch gradient descent algorithm.

Listing 1-3 implements the function for linear regression with the mini-batch gradient descent algorithm in Python [8]. These functions are then applied to a simulated dataset.

Listing 1-3. Implementation of Linear Regression with Mini-Batch Gradient Descent

```
###############################################################
# Import the libraries
import numpy as np
import matplotlib.pyplot as plt
import pandas as pd
from pandas import read_csv

np.random.seed(111)

# creating data
mean = np.array([6.0, 7.0])
covariance = np.array([[1.0, 0.94], [0.95, 1.2]])
df = np.random.multivariate_normal(mean, covariance, 500)

df.shape
# output- (10000, 2)

df
#Output-
#array([[6.97833668, 8.30414776],
#       [4.66509294, 5.31806407],
```

```
#         [6.88804517, 7.71734893],
#         ...,
#         [5.75153443, 6.73145512],
#         [4.93357924, 6.72570148],
#         [7.31794626, 8.4076224 ]])

X = df[:,:-1]
Y = df[:,-1]
# A column with all 1's is added
X_new = np.c_[np.ones((500, 1)), X]

# Random initialization of the estimate "reg_coef"
np.random.seed(333)
reg_coef = np.random.randn(2,1)

# here, size_batch_mini= minibatch size
# lr= learning rate
#  max_iters= number of batches used
lr=0.01
num = 100
max_iters = 100
size_batch_mini = 50

t0, t1 = 400, 1200
def lrs(step):
    return t0 / (step + t1)

reg_coef_all = []

step = 0
for j in range(max_iters):
    batches_index  = np.random.permutation(num)
    X_batches = X_new[batches_index]
    y_batches = Y[batches_index]
```

```
    for i in range(0, num, size_batch_mini):
        step += 1
        Yi = y_batches[i:i+size_batch_mini]
        Xi = X_batches[i:i+size_batch_mini]
        # compute the gradient
        gradient = 2/size_batch_mini * Xi.T.dot(Xi.dot
        (reg_coef) - Yi)
        lr = lrs(step)
        # update
        reg_coef = reg_coef - lr * gradient
        reg_coef_all.append(reg_coef)

# Output
reg_coef
reg_coef_all
###############################################################
```

Similarly, the linear regression method can be converted to an online learning method by using the SGD method. I urge you to implement the function for the linear regression method with SGD using the ideas discussed so far. You can refer to Listing 1-3, but make sure that the update step happens for each observation.

The scikit-learn library has an SGDRegressor() function to apply linear regression with SGD. Listing 1-4 is a code snippet for applying linear regression using SGD on a simulated dataset.

Listing 1-4. Stochastic Gradient Descent Implementation Using scikit-learn

```
###############################################################
# Import relevant libraries
import numpy as np
import pandas as pd
from sklearn.model_selection import train_test_split
```

```
from sklearn.linear_model import SGDRegressor
from sklearn.metrics import mean_squared_error, r2_score

np.random.seed(111)

# creating the dataset
mean = np.array([6.0, 7.0])
covariance = np.array([[1.0, 0.94], [0.95, 1.2]])
df = np.random.multivariate_normal(mean, covariance, 10000)

df.shape
# output- (10000, 2)
df
#Output-
#array([[6.97833668, 8.30414776],
#        [4.66509294, 5.31806407],
#        [6.88804517, 7.71734893],
#        ...,
#        [5.75153443, 6.73145512],
#        [4.93357924, 6.72570148],
#        [7.31794626, 8.4076224 ]])
X = df[:,:-1]
Y = df[:,-1]

## Split into train and test
train_X, test_X, train_y, test_y = train_test_split(X, Y,
test_size=0.20, random_state=333)

### Use the SGD regressor
mod = SGDRegressor()

### Fit the model
mod.fit(train_X, train_y)
```

```
#SGDRegressor(alpha=0.0001, average=False, early_
stopping=False, epsilon=0.1,
#               eta0=0.01, fit_intercept=True, l1_ratio=0.15,
#               learning_rate='invscaling', loss='squared_loss',
                max_iter=1000,
#               n_iter_no_change=5, penalty='l2', power_t=0.25,
                random_state=None,
#               shuffle=True, tol=0.001, validation_fraction=0.1,
                verbose=0,
#               warm_start=False)

### Print the coefficient and intercept values
print("Coefficients: \n", mod.coef_)
#Coefficients: [0.95096598]

print("Intercept", mod.intercept_)
#Intercept [1.17712595]

### Predict on the test data
pred = mod.predict(test_X)
# calculating the prediction error
error = np.sum(np.abs(test_y - pred) / test_y.shape[0])
print("Mean absolute error (MAE) = ", error)
# Mean absolute error (MAE) =  0.4493164833335055
##############################################################
```

This section explained incremental learning and online learning algorithms and implemented a couple of these algorithms in a linear regression context using Python. In the next section, you are introduced to the scikit-multiflow.

Introduction to the Scikit-Multiflow Framework

scikit-multiflow is a framework for the analysis of streaming data in Python [9]. It is an open source software distributed under the BSD license. There was a need for an open source machine learning library in Python for streaming data, similar to what the scikit-learn library [10] has been doing for static/batch data for years. The scikit-multiflow framework is built upon other famous open source frameworks, such as scikit-learn, MOA [11], and MEKA [12].

MOA is among the oldest and most popular open source software for data stream mining. It contains a collection of streaming data algorithms for supervised (classification, regression, etc.) and unsupervised learning (clustering, etc.). MOA is developed in Java.

MEKA is an open source software for multilabel learning and evaluation. scikit-multiflow complements scikit-learn (whose primary focus is on batch learning) and helps with machine learning in streaming data research.

The scikit-multiflow website (https://scikit-multiflow.github.io/) provides official documentation and the user guide.

The API reference documentation is available at https://scikit-multiflow.readthedocs.io/en/stable/api/api.html.

The scikit-multiflow package source code is publicly available at https://github.com/scikit-multiflow/scikit-multiflow.

numpy and cython are two dependencies that need to be installed before you install scikit-multiflow.

scikit-multiflow only works only on Python 3.5+ versions.

To install scikit-multiflow from the Python Package Index (PyPI), use the following command in the command prompt/terminal.

```
$ pip install -U scikit-multiflow
```

scikit-multiflow can also be installed from conda-forge using the following command.

```
$ conda install -c conda-forge scikit-multiflow
```

This book uses the version 0.5.3 package, released in June 2020.

In the next section, you learn about the various streaming data generators available in the scikit-multiflow framework.

Streaming Data Generators

The scikit-multiflow framework contains many streaming data generators to create simulated data streams for experimental and research-related work. One of the popular streaming data generators is the hyperplane generator. A hyperplane in a p-dimensional space is the set of points (x) that satisfy the following.

$$\sum_{j=1}^{p} w_j x_j = w_0$$

x_j is the jth coordinate of x. w is the weight [13].

For the target variable, the following types of observations are labeled as positive.

$$\sum_{j=1}^{p} w_j x_j >= w_0$$

The following types of observations are labeled as negative.

$$\sum_{j=1}^{p} w_j x_j < w_0$$

Hyperplanes are useful for simulating time-changing concepts as the orientation. The position of the hyperplane can be changed by adjusting the relative size of the weight. Listing 1-5 generates a data stream using scikit-multiflow comprising ten features and a target variable. It generates 10,000 observations before storing the simulated data in a CSV file. Figure 1-5 is a snapshot of the generated dataset.

Listing 1-5. Streaming Data Generation Using the Hyperplane Generator

```
############################################################
# Import the relevant libraries
from skmultiflow.data import HyperplaneGenerator
import pandas as pd
import numpy as np

# Apply the HyperplaneGenerator() function
create = HyperplaneGenerator(random_state = 888, n_features= 10,
noise_percentage = 0)
create.prepare_for_use()
X , Y = create.next_sample(10000)
data = pd.DataFrame(np.hstack((X, np.array([Y]).T)))

data.shape
# output- (10000, 11)

print(data.head())
# Output:
#           0          1          2          3    ...           7
       8          9         10
#0   0.899478   0.248365   0.030172   0.072447    ...    0.633460
0.283253   0.365369   1.0
#1   0.705171   0.648509   0.040909   0.211732    ...    0.026095
0.446087   0.239105   0.0
```

```
#2   0.091587   0.977452   0.411501   0.458305   ...   0.181444
0.303406   0.174454   0.0
#3   0.635272   0.496203   0.014126   0.627222   ...   0.517752
0.570683   0.546333   1.0
#4   0.450078   0.876507   0.537356   0.495684   ...   0.606895
0.217841   0.912944   1.0
#
#[5 rows x 11 columns]

# Store the data in csv
data.to_csv('data_stream_hyperplane.csv', index=False)
##############################################################
```

Index	0	1	2	3	4	5	6	7	8	9	10
0	0.899478	0.248365	0.0301718	0.0724471	0.074164	0.55843	0.916847	0.63346	0.283253	0.365369	1
1	0.705171	0.648509	0.0409088	0.211732	0.00148992	0.138972	0.211825	0.0208949	0.446887	0.239185	0
2	0.0915871	0.977452	0.411501	0.458305	0.525909	0.294416	0.972116	0.181444	0.383406	0.174454	0
3	0.635272	0.496203	0.0141264	0.627222	0.634975	0.100141	0.839847	0.517752	0.570683	0.546333	1
4	0.450078	0.876507	0.537356	0.495684	0.418272	0.851606	0.306053	0.606895	0.217841	0.912944	1
5	0.856398	0.180293	0.948519	0.384163	0.251388	0.967466	0.77048	0.446852	0.198139	0.839823	1
6	0.578489	0.856236	0.357642	0.476238	0.545923	0.795083	0.144624	0.0180292	0.385322	0.906146	0
7	0.00179492	0.214237	0.28306	0.594816	0.42567	0.522077	0.780824	0.207883	0.768618	0.595955	0
8	0.175171	0.454754	0.461879	0.797812	0.0510449	0.424061	0.123532	0.918123	0.219467	0.0592571	0
9	0.563733	0.599512	0.166961	0.915744	0.226674	0.0908589	0.689444	0.391173	0.320342	0.722016	1
10	0.753456	0.889861	0.338064	0.197249	0.864116	0.962722	0.739514	0.674087	0.377465	0.794774	1
11	0.211648	0.824925	0.0736159	0.636996	0.303908	0.8696185	0.681359	0.56667	0.783518	0.0918758	0
12	0.625633	0.990847	0.393813	0.214151	0.693516	0.8356316	0.462469	0.537626	0.522928	0.464937	0
13	0.110446	0.92658	0.126959	0.846598	0.8312274	0.394362	0.582441	0.254377	0.298248	0.217418	0
14	0.25868	0.915166	0.0860515	0.723262	0.697547	0.957633	0.595555	0.194041	0.536536	0.587583	0
15	0.569669	0.16752	0.954324	0.948005	0.704714	0.815857	0.080253	0.33374	0.017156	0.698215	1
16	0.578307	0.241308	0.124323	0.156636	0.213772	0.705631	0.736411	0.2066	0.388298	0.376257	0

Figure 1-5. *Snapshot of the dataset generated using the hyperplane generator*

Another popular streaming data generator is the Agarwal generator. The Agarwal stream generator is one of the earliest data sources for training decision tree learners [14].

Listing 1-6 generates a data stream using the Agarwal generator. A snapshot of this dataset is shown in Figure 1-6. The dataset has ten columns. The rightmost column is the binary target variable (0 or 1).

There are nine features in the dataset. The target variable represents whether a loan should be approved or not. Some of the features or predictor variables in this dataset are age, salary, education level, loan amount, the value of the house, zip code, commission, the carmaker, and the number of years the house is owned.

Listing 1-6. Agarwal Streaming Data Generation

```
###############################################################
# Import the relevant libraries
from skmultiflow.data import HyperplaneGenerator
import pandas as pd
import numpy as np

# Apply the HyperplaneGenerator() function
create = HyperplaneGenerator(random_state = 888, n_features=
10, noise_percentage = 0)
create.prepare_for_use()
X , Y = create.next_sample(10000)
data = pd.DataFrame(np.hstack((X, np.array([Y]).T)))

data.shape
# output- (10000, 11)

print(data.head())
# Output:
#          0          1          2          3    ...          7
       8          9         10
#0   0.899478   0.248365   0.030172   0.072447   ...   0.633460
0.283253   0.365369   1.0
#1   0.705171   0.648509   0.040909   0.211732   ...   0.026095
0.446087   0.239105   0.0
#2   0.091587   0.977452   0.411501   0.458305   ...   0.181444
0.303406   0.174454   0.0
```

```
#3   0.635272  0.496203  0.014126  0.627222  ...  0.517752
0.570683  0.546333  1.0
#4   0.450078  0.876507  0.537356  0.495684  ...  0.606895
0.217841  0.912944  1.0
#
#[5 rows x 11 columns]

# Store the data in csv
data.to_csv('data_stream_hyperplane.csv', index=False)
###############################################################
```

Index	0	1	2	3	4	5	6	7	8	9
0	90627.8	0	33	3	19	6	329198	20	24151.8	0
1	33588.9	17307.8	72	4	9	7	112096	29	315025	0
2	24375.1	12426.9	39	3	19	3	678134	4	363159	0
3	82949.7	0	68	0	2	7	216270	2	35758.5	0
4	149424	0	52	0	5	3	447136	29	98440.4	1
5	49871.6	25601.8	80	2	14	1	1.13657e+06	2	84403.2	0
6	57108.8	79705.8	63	3	10	2	565482	13	321814	0
7	91069.4	84896.6	73	0	16	6	247338	6	20998.1	0
8	42763.2	84896.6	78	0	1	8	91733	4	228067	0
9	51632	62872.7	75	1	17	5	493089	21	491337	0
10	84608.9	0	62	2	2	2	740006	21	212440	0
11	35519.9	18750.2	74	0	3	8	136630	22	199485	0
12	50307	67444.1	72	1	10	2	647640	28	41062.6	0
13	103591	0	79	4	5	4	346815	11	459851	0
14	45387.7	79515.5	38	0	11	3	695364	13	463142	0
15	61859.9	83856.7	24	4	1	1	746747	22	395593	0
16	35492.8	77650.6	33	4	19	4	391551	20	180473	0
17	145621	0	40	2	2	1	1.18524e+06	8	277660	1
18	123178	0	24	4	13	5	400450	23	490383	0

Figure 1-6. *Snapshot of the dataset generated using the Agarwal generator*

The following are some of the many other streaming data generators available in the scikit-multiflow framework.

- LED stream generator and LED stream generator with concept drift

- Sine generator and anomaly sine generator

23

- Mixed data stream generator

- Random Radial Basis Function stream generator and
  Random Radial Basis Function stream generator with
  concept drift

- SEA stream generator

- Random tree stream generator

- STAGGER concepts stream generator

- Waveform stream generator

- Regression generator

This section discussed the various streaming data generators. In the
next section, you learn how to create a data stream from a CSV file.

Create a Data Stream from a CSV file

The FileStream method in a scikit-multiflow data module creates a
data stream from a file source. As of September 2020, only CSV files
were supported. You can take any static/batch dataset and convert it to
streaming data.

As a demonstration, let's look at an absenteeism at the workplace
dataset [15] that is publicly available on the UCI Machine Learning
Repository [16]. The dataset is in CSV format. It contains various attributes:
employee ID, the reason for absence, month/day/season of absence,
employee age, average workload per day, education, body mass index,
absenteeism time in hours, and more. This dataset has 21 attributes and
740 observations.

Figure 1-7 is a snapshot of the dataset containing 19 rows and 11 columns.
Listing 1-7 uses the FileStream method to load the absenteeism-at-work CSV
file and then convert it to a data stream. The next_sample() function retrieves

the next sample. Also, the built-in functions, has_more_samples() and n_ remaining_samples(), tell you if there are more samples and the number of remaining samples, respectively.

ID	Reason for absence	Month of absence	Day of the week	Seasons	nsportation expen	Distance from Residence to Work	Service time	Age	Work load Average/day	Hit target
11	26	7	3	1	289	36	13	33	239.554	97
36	0	7	3	1	118	13	18	50	239.554	97
3	23	7	4	1	179	51	18	38	239.554	97
7	7	7	5	1	279	5	14	39	239.554	97
11	23	7	5	1	289	36	13	33	239.554	97
3	23	7	6	1	179	51	18	38	239.554	97
10	22	7	6	1	361	52	3	28	239.554	97
20	23	7	6	1	260	50	11	36	239.554	97
14	19	7	2	1	155	12	14	34	239.554	97
1	22	7	2	1	235	11	14	37	239.554	97
20	1	7	2	1	260	50	11	36	239.554	97
20	1	7	3	1	260	50	11	36	239.554	97
20	11	7	4	1	260	50	11	36	239.554	97
3	11	7	4	1	179	51	18	38	239.554	97
3	23	7	4	1	179	51	18	38	239.554	97
24	14	7	6	1	246	25	16	41	239.554	97
3	23	7	6	1	179	51	18	38	239.554	97
3	21	7	2	1	179	51	18	38	239.554	97
6	11	7	5	1	189	29	13	33	239.554	97

Figure 1-7. *Snapshot of the absenteeism at workplace dataset (all columns are not shown)*

Listing 1-7. Creating data stream from a CSV file

```
##############################################################
# Import the relevant library
from skmultiflow.data.file_stream import FileStream

# Setup the data stream
data_stream= FileStream('./Absenteeism_at_work.csv')

# Retrieving one sample
data_stream.next_sample()
# Output-
#(array([[ 11.  ,   26.  ,    7.  ,    3.  ,    1.  ,  289.  ,   36.  ,
#           13.  ,   33.  ,  239.554,   97.  ,    0.  ,    1.  ,    2.  ,
#            1.  ,    0.  ,    1.  ,   90.  ,  172.  ,   30.  ]]), array([4]))
```

```
 # Retrieving 5 samples
data_stream.next_sample(5)
# Output-
#(array([[ 36.   ,   0.   ,   7.   ,   3.   ,   1.   , 118.   ,  13.   ,
#          18.   ,  50.   , 239.554,  97.   ,   1.   ,   1.   ,   1.   ,
#           1.   ,   0.   ,   0.   ,  98.   , 178.   ,  31.   ],
#        [  3.   ,  23.   ,   7.   ,   4.   ,   1.   , 179.   ,  51.   ,
#          18.   ,  38.   , 239.554,  97.   ,   0.   ,   1.   ,   0.   ,
#           1.   ,   0.   ,   0.   ,  89.   , 170.   ,  31.   ],
#        [  7.   ,   7.   ,   7.   ,   5.   ,   1.   , 279.   ,   5.   ,
#          14.   ,  39.   , 239.554,  97.   ,   0.   ,   1.   ,   2.   ,
#           1.   ,   1.   ,   0.   ,  68.   , 168.   ,  24.   ],
#        [ 11.   ,  23.   ,   7.   ,   5.   ,   1.   , 289.   ,  36.   ,
#          13.   ,  33.   , 239.554,  97.   ,   0.   ,   1.   ,   2.   ,
#           1.   ,   0.   ,   1.   ,  90.   , 172.   ,  30.   ],
#        [  3.   ,  23.   ,   7.   ,   6.   ,   1.   , 179.   ,  51.   ,
#          18.   ,  38.   , 239.554,  97.   ,   0.   ,   1.   ,   0.   ,
#           1.   ,   0.   ,   0.   ,  89.   , 170.   ,  31.   ]]),
# array([0, 2, 4, 2, 2]))

data_stream.has_more_samples()
# Output-
# True

data_stream.n_remaining_samples()
# Output-
# 734
################################################################
```

Summary

In this chapter, you learned about streaming data, the various challenges associated with it, real-world business applications, windowing techniques, and incremental and online learning algorithms. You were also introduced to the scikit-multiflow framework in Python.

The next chapter discusses concept drift detection and anomaly detection algorithms and their implementation in various datasets using the scikit-multiflow framework.

References

[1]   "Big data: The next frontier for innovation, competition, and productivity." McKinsey Global Institute, 2011.

[2]   B. Ellis. *Real-Time Analytics: Techniques to Analyze and Visualize Streaming Data.* Wiley, 2014.

[3]   J. Gama. *Knowledge Discovery from Data Streams.* Chapman & Hall/CRC, 2010.

[4]   A. Hutchison. "Facebook Reaches 2.38 Billion Users, Beats Revenue Estimates in Latest Update." *Social Media Today*, April 24, 2019.

[5]   Stefan Büttcher, Charles L. A. Clarke, and Gordon V. Cormack. *Information Retrieval: Implementing and Evaluating Search Engines.* MIT Press, 2010.

[6]   Stratos Mansalis, Eirini Ntoutsi, Nikos Pelekis, and Yannis Theodoridis . "An evaluation of data stream clustering algorithms." *Statistical Analysis and Data Mining*, June 25, 2018.

[7]   A. Laha and S. Putatunda. "Real-time location
        prediction with taxi-gps data streams,"
        *Transportation Research Part C: Emerging
        Technologies*, July, 2018.

[8]   A. Géron. *Hands-on Machine Learning with
        Scikit-Learn and TensorFlow: Concepts, Tools, and
        Techniques to Build Intelligent Systems*. O'Reilly
        Media, 2017.

[9]   Jacob Montiel, Jesse Read, Albert Bifet, and Talel
        Abdessalem. "Scikit-Multiflow: A Multi-Output
        Streaming Framework." *The Journal of Machine
        Learning Research*, 2018.

[10]   Fabian Pedregosa, Gaël Varoquaux, Alexandre
        Gramfort, Vincent Michel, Bertrand Thirion,
        Olivier Grisel, Mathieu Blondel, Peter Prettenhofer,
        Ron Weiss, Vincent Dubourg, et al. "Scikit-learn:
        Machine Learning in Python." *Journal of Machine
        Learning Research*. October 12, 2011.

[11]   Albert Bifet, Ricard Gavalda, Geoff Holmes, and Bernhard
        Pfahringer. *Machine Learning for Data Streams with
        Practical Examples in MOA*. MIT Press, 2018.

[12]   Jesse Read, Peter Reutemann, Bernhard Pfahringer,
        and Geoff Holmes. "MEKA: A multi- label/multi-
        target extension to WEKA." *Journal of Machine
        Learning Research*. January, 2016.

[13]   G. Hulten, L. Spencer, and P. Domingos. *Mining
        Time-Changing Data Streams*. ACM Press, 2001.

[14]   Rakesh Agrawal, Tomasz Imielinksi, and Arun Swami. "Database Mining: A Performance Perspective", IEEE Transactions on Knowledge and Data Engineering, December 1993.

[15]   A. Martiniano, R. P. Ferreira, R. J. Sassi, and C. Affonso. "Application of a neuro fuzzy network in prediction of absenteeism at work." IEEE Conference on Information Systems and Technologies (CISTI), 2012.

[16]   D. Dua and C. Graff. UCI Machine Learning Repository (http://archive.ics.uci.edu/ml), 2019.

CHAPTER 2

Concept Drift Detection in Data Streams

In the last chapter, you were introduced to streaming data, its applications, windowing techniques, and incremental and online learning algorithms. You were also introduced to the scikit-multiflow framework in Python and the various streaming data generators. This chapter discusses algorithms for concept drift detection in a streaming data context. It explores relevant techniques and how to implement them using the scikit-multiflow framework in Python.

Note This chapter applies some of the scikit-multiflow framework's (version 0.5.3) drift detection module functions. Also, I refer to the scikit-multiflow framework's API reference document for the codes provided in this chapter (`https://scikit-multiflow.readthedocs.io/en/stable/api/api.html`).

Concept Drift

Chapter 1 discussed streaming data's various challenges. One of the major challenges is concept drift, in which the distribution of data streams may change over time. A particular pattern present in the past data may become a different kind of pattern in current data. So, any batch learner model trained on past data becomes outdated due to concept drift.

© Sayan Putatunda 2021
S. Putatunda, *Practical Machine Learning for Streaming Data with Python*,
https://doi.org/10.1007/978-1-4842-6867-4_2

In machine learning, *concept* refers to a variable or a quantity that a machine learning model is trying to predict. *Concept drift* refers to the phenomenon in which the target concept's statistical properties change over time [1].

The changes in data streams or concept drift patterns can be either gradual or abrupt [2]. Abrupt changes in data streams, or abrupt concept drift, mean the sudden change in the characteristics of the data, such as a change in mean, change in variance, and so forth. It is important to detect these changes because they have many practical applications in quality control, systems monitoring, fault detection, and more. However, gradual changes in data streams or gradual concept drift are referred to as the *change in the data distribution* over a much longer time than that of the abrupt changes. Gradual changes are relatively more difficult to identify.

The other types of data streams or concept drift changes are *incremental concept drift* and *recurring concept drift* [3]. In incremental concept drift, the older patterns in data streams incrementally change to newer patterns over time; however, the time taken is much shorter than that of the gradual concept drift. In recurring concept drift, you see specific characteristics of older data streams reoccurring after a time.

There are various types of concept drift detection algorithms. This chapter focuses on concept drift detection algorithms such as (a) the *adaptive windowing method* for concept drift detection (ADWIN), (b) the Drift Detection Method (DDM), (c) the Early Drift Detection Method (EDDM), (d) Drift Detection Method based on Hoeffding's bound with moving average-test (HDDM_A), (e) Drift Detection Method based on Hoeffding's bound with moving weighted average-test (HDDM_W), and (f) drift detection using the Page-Hinkley method.

Adaptive Windowing Method for Concept Drift Detection

In the adaptive windowing method for concept drift detection (ADWIN), a sliding window of dynamic size is used; that is, the window's size is not fixed, but it is computed online based on the rate of change observed from the data in the window [4]. I already discussed the idea of sliding windows in Chapter 1, where a window is maintained with the most recent data points in a data stream, and the older data points are continuously discarded. The data points within the window get the data stream's updated statistics, detect concept drift, and update the model.

It is generally recommended to use windows with a variable size. The idea is to keep data points or observations for as long as possible since the change time scale is unknown to the users [4]. The ADWIN algorithm increases the window size if it doesn't see any change in the data stream, and reduces the window size when a change is detected in the data stream. The idea is to track the mean of the data points in window W. The null hypothesis is that there is no change in the mean value inside window W. The older data points in the window are dropped when there is enough evidence that the mean value of the new data points differs from that of the rest of the window [5].

To dive further into the implementation of the ADWIN algorithm, a statistical test is performed to check if the mean value in two windows (W_1 and W_2) differs by more than a threshold. W_1 and W_2 are the subwindows of the main window, W. If the subwindows' means are significantly different, the older data points of window W are dropped. However, if the mean of subwindows are not significantly different, it is assumed that there is no change in the distribution of the data stream [4].

Listing 2-1 illustrates concept drift detection using ADWIN on a simulated data stream. The drift_detection module is used in the scikit-multiflow framework. Listing 2-1 simulates a data stream of size 1000 from a standard normal distribution and then changes the data concept from index 599 onward (i.e., replaces values with integers between 5 and 8), as shown in Figure 2-1. It then calls the ADWIN() object and uses the detected_change() function to identify concept drift in the data stream. The final output is shown in Listing 2-2.

Listing 2-1. Illustration for Concept Drift Detection Using ADWIN in scikit-multiflow

```
###############################################################

# Import the relevant libraries
import numpy as np
from skmultiflow.drift_detection.adwin import ADWIN

# set seed for reproducibility
np.random.seed(123)

# Simulate a data stream of size 1000 from a Standard normal
distribution
stream = np.random.randn(1000)

stream[:10]
# Output:
#array([-1.0856306 ,  0.99734545,  0.2829785 , -1.50629471,
        -0.57860025,
#        1.65143654, -2.42667924, -0.42891263,  1.26593626,
        -0.8667404 ])
```

```
# Data concept are changed from index 599 to 999
for j in range(599, 1000):
    stream[j] = np.random.randint(5, high=9)

# Initialize the ADWIN object
A = ADWIN(delta=0.002)

# Stream elements are added to ADWIN and checking whether drift
occured
for j in range(1000):
    A.add_element(stream[j])
    if A.detected_change():
        print('Concept Drift detected in data: ' +
        str(stream[j]) + ' - at index: ' + str(j))
##################################################################
```

Listing 2-2. Output for Concept Drift Detection Using ADWIN

```
##################################################################
### Output:
Concept Drift detected in data: 8.0 - at index: 607
Concept Drift detected in data: 5.0 - at index: 639
Concept Drift detected in data: 6.0 - at index: 671
##################################################################
```

	0
587	-1.66048
588	0.588686
589	1.33319
590	2.55985
591	0.0349026
592	0.23265
593	1.6066
594	0.168722
595	0.275342
596	-0.630618
597	-1.39438
598	0.912688
599	5
600	6
601	7
602	8
603	8
604	5
605	8
606	5
607	8

Figure 2-1. *The data stream with changed concept from index 599 onward*

Drift Detection Method

The Drift Detection Method (DDM) was first proposed by Gama et al. [6]. The DDM method can be applied with any learning algorithm; that is, with either incremental/online learning algorithms or batch learning algorithms. The idea is to keep track of the learning algorithm's error rate. In a sample of n examples, the number of errors from the learning algorithm is a random variable from Bernoulli trials. You can use the binomial distribution to model the number of errors in n examples. At instant j, the error rate is denoted by p_j with a standard deviation of $\sigma_j = sqrt(p_j(1-p_j)/j)$.

The DDM method keeps a record of the p_{min} and σ_{min} while training the learning algorithm [6]. If $p_j + \sigma_j >= p_{min} + 2 * \sigma_{min}$, then a warning is declared, which means that there is a change in the data stream. But if $p_j + \sigma_j >= p_{min} + 3 * \sigma_{min}$, a change in the data stream is declared [5].

Listing 2-3 illustrates concept drift detection using DDM. It simulates a data stream of size 1000 from a standard normal distribution and then changes the data concept from index 299 to 599 (i.e., replace values with integers between 5 and 9). It then calls the DDM() object and uses the detected_change() and detected_warning_zone() functions to identify concept drift in data streams and provide a warning about a change in the data streams, respectively. The final output is shown in Listing 2-4.

Listing 2-3. Illustration for Concept Drift Detection Using DDM in scikit-multiflow

```
###############################################################
# Import the relevant libraries
import numpy as np
from skmultiflow.drift_detection import DDM

# set seed for reproducibility
np.random.seed(123)
```

```
# Simulate a data stream of size 1000 from a Standard normal
  distribution
stream = np.random.randn(1000)
stream[:10]
## Output-
#array([-1.0856306 ,  0.99734545,  0.2829785 , -1.50629471,
        -0.57860025,
#        1.65143654, -2.42667924, -0.42891263,  1.26593626,
        -0.8667404 ])
# Data concepts are changed from index 299 to 599
for j in range(299, 600):
    stream[j] = np.random.randint(5, high=9)
# Initialize the DDM object
d2m = DDM()
# Stream elements are added to DDM and checking whether drift
  occurred
for j in range(1000):
    d2m.add_element(stream[j])
    if d2m.detected_change():
        print('Concept drift detected in data: ' +
        str(stream[j]) + ' - at index: ' + str(j))
    if d2m.detected_warning_zone():
        print('Warning detected in data: ' + str(stream[j]) +
        ' - at index: ' + str(j))
################################################################
```

Listing 2-4. Output for Concept Drift and Warning Zones Detected Using DDM

```
################################################################
### Output:
Concept drift detected in data: 1.0693159694243486 - at index: 55
Concept drift detected in data: 2.0871133595881854 - at index: 88
```

```
Concept drift detected in data: 0.8123413299768204 - at index: 126
Warning detected in data: 1.3772574828673068 - at index: 158
Warning detected in data: -0.1431759743261871 - at index: 159
Warning detected in data: 0.02031599823462459 - at index: 160
Warning detected in data: -0.19396387055266243 - at index: 161
Warning detected in data: 0.13402679274666512 - at index: 162
Warning detected in data: 0.7044740740436035 - at index: 163
Concept drift detected in data: 0.6656534379123312 - at index: 164
Concept drift detected in data: 8.0 - at index: 302
###################################################################
```

Early Drift Detection Method

The Early Drift Detection Method (EDDM) is an improvement of the
DDM method, especially for detecting slow, gradual changes in a data
stream. In the DDM method discussed earlier, the idea was to keep track
of the number of errors in the learning algorithms, whereas in EDDM, the
mean distance of two errors is tracked [7]. Thus, EDDM stores the average
distance between two errors (p'_j and its standard deviation σ'_j) and the
maximum values p'_{max} and σ'_{max}. Similar to that of DDM, based on certain
thresholds, either a concept drift or a warning zone is declared. Please
refer to Baena-Garcia et al. [7] for a detailed review of the EDDM.

Concept drift detection using EDDM can be implemented in the same
way as in Listing 2-3. Instead of using the DDM() method, however, you
need to apply the EDDM() method. You are encouraged to change the
distribution of the data stream and note the output.

Drift Detection Using HDDM_A and HDDM_W

Drift detection using moving average-test (HDDM_A) is based on
Hoeffding's bound [10]. Hoeffding's inequality does not assume a

probability function; it assumes independent and bounded random variables [8]. HDDM_A performs an *A-test*, in which given a sequence of random variables, changes in the population mean can be detected by monitoring the difference between the moving averages [8]. The HDDM_A method is implemented in scikit-multiflow. The implementation of HDDM_A is based on MOA [9].

Listing 2-5 illustrates concept drift detection and warning zone detection using the HDDM_A method in scikit-multiflow. It simulates a data stream of size 1000 from a binomial distribution and changes the data concepts from index 299 to 499. The final output is shown in Listing 2-6.

Listing 2-5. Illustration for Concept Drift Detection Using HDDM_A in scikit-multiflow

```
###############################################################
# Import the relevant libraries
 import numpy as np
from skmultiflow.drift_detection.hddm_a import HDDM_A

# Initialize the HDDM_A object
HA = HDDM_A()

# set seed for reproducibility
np.random.seed(123)

# Simulate a data stream of size 1000 from a binomial
distribution
# here, n= number of trials, p= probability of each trial
n, p = 10, 0.6
stream = np.random.binomial(n, p, 1000)
stream[:10]
# Output- array([5, 7, 7, 6, 5, 6, 3, 5, 6, 6])
```

```
# Data concept are changed from index 299 to 499
for j in range(299, 500):
    stream[j] = np.random.randint(5, high=9)

# Stream elements are added to DDM and checking whether drift
  occured
for j in range(1000):
    HA.add_element(stream[j])
    if HA.detected_change():
        print('Concept drift detected in data: ' +
        str(stream[j]) + ' - at index: ' + str(j))
    if HA.detected_warning_zone():
        print('Warning detected in data: ' + str(stream[j]) +
        ' - at index: ' + str(j))
####################################################################
```

Listing 2-6. Output for Concept Drift and Warning Zones Detected Using HDDM_A

```
####################################################################
### Output:
Concept drift detected in data: 8 - at index: 13
Warning detected in data: 7 - at index: 26
Concept drift detected in data: 7 - at index: 27
Concept drift detected in data: 8 - at index: 65
Warning detected in data: 8 - at index: 70
Concept drift detected in data: 9 - at index: 74
Concept drift detected in data: 8 - at index: 78
Concept drift detected in data: 7 - at index: 92
Concept drift detected in data: 8 - at index: 96
Concept drift detected in data: 8 - at index: 103
Warning detected in data: 7 - at index: 111
Concept drift detected in data: 7 - at index: 112
```

Concept drift detected in data: 10 - at index: 118
Concept drift detected in data: 7 - at index: 121
Concept drift detected in data: 7 - at index: 126
Concept drift detected in data: 8 - at index: 131
Concept drift detected in data: 8 - at index: 147
Concept drift detected in data: 8 - at index: 150
Warning detected in data: 8 - at index: 163
Concept drift detected in data: 8 - at index: 171
Concept drift detected in data: 8 - at index: 175
Concept drift detected in data: 7 - at index: 190
Concept drift detected in data: 9 - at index: 194
Concept drift detected in data: 9 - at index: 196
Concept drift detected in data: 8 - at index: 199
Concept drift detected in data: 8 - at index: 206
Concept drift detected in data: 8 - at index: 209
Concept drift detected in data: 8 - at index: 225
Warning detected in data: 8 - at index: 231
Concept drift detected in data: 10 - at index: 240
Concept drift detected in data: 9 - at index: 258
Concept drift detected in data: 10 - at index: 272
Concept drift detected in data: 9 - at index: 279
Concept drift detected in data: 9 - at index: 281
Concept drift detected in data: 6 - at index: 290
Concept drift detected in data: 7 - at index: 292
Concept drift detected in data: 8 - at index: 294
Concept drift detected in data: 8 - at index: 298
Concept drift detected in data: 8 - at index: 365
Warning detected in data: 8 - at index: 380
Concept drift detected in data: 8 - at index: 381
Concept drift detected in data: 8 - at index: 411
Concept drift detected in data: 8 - at index: 417
Concept drift detected in data: 8 - at index: 423

```
Concept drift detected in data: 8 - at index: 445
Warning detected in data: 8 - at index: 451
Warning detected in data: 7 - at index: 452
Concept drift detected in data: 8 - at index: 468
Concept drift detected in data: 8 - at index: 497
Warning detected in data: 8 - at index: 503
Concept drift detected in data: 9 - at index: 547
Warning detected in data: 9 - at index: 561
Warning detected in data: 9 - at index: 566
Warning detected in data: 8 - at index: 567
Warning detected in data: 5 - at index: 568
Warning detected in data: 6 - at index: 569
Warning detected in data: 6 - at index: 570
Warning detected in data: 7 - at index: 571
Warning detected in data: 7 - at index: 572
Concept drift detected in data: 9 - at index: 586
Warning detected in data: 8 - at index: 593
Concept drift detected in data: 9 - at index: 598
Warning detected in data: 6 - at index: 604
Concept drift detected in data: 6 - at index: 605
Concept drift detected in data: 8 - at index: 619
Concept drift detected in data: 6 - at index: 622
Concept drift detected in data: 7 - at index: 628
Concept drift detected in data: 8 - at index: 640
Concept drift detected in data: 9 - at index: 644
Concept drift detected in data: 8 - at index: 655
Concept drift detected in data: 8 - at index: 660
Concept drift detected in data: 7 - at index: 662
Concept drift detected in data: 8 - at index: 670
Concept drift detected in data: 8 - at index: 688
Concept drift detected in data: 7 - at index: 697
Concept drift detected in data: 7 - at index: 701
```

Concept drift detected in data: 9 - at index: 703
Concept drift detected in data: 8 - at index: 705
Concept drift detected in data: 9 - at index: 716
Concept drift detected in data: 9 - at index: 719
Concept drift detected in data: 8 - at index: 723
Concept drift detected in data: 7 - at index: 750
Warning detected in data: 8 - at index: 756
Concept drift detected in data: 7 - at index: 772
Concept drift detected in data: 7 - at index: 775
Concept drift detected in data: 8 - at index: 782
Concept drift detected in data: 8 - at index: 788
Concept drift detected in data: 9 - at index: 802
Concept drift detected in data: 7 - at index: 806
Concept drift detected in data: 6 - at index: 808
Concept drift detected in data: 7 - at index: 824
Concept drift detected in data: 8 - at index: 826
Warning detected in data: 7 - at index: 830
Warning detected in data: 7 - at index: 836
Concept drift detected in data: 7 - at index: 837
Concept drift detected in data: 6 - at index: 839
Concept drift detected in data: 9 - at index: 847
Concept drift detected in data: 8 - at index: 853
Concept drift detected in data: 6 - at index: 855
Concept drift detected in data: 9 - at index: 864
Concept drift detected in data: 8 - at index: 867
Concept drift detected in data: 8 - at index: 870
Concept drift detected in data: 9 - at index: 874
Concept drift detected in data: 8 - at index: 882
Concept drift detected in data: 7 - at index: 894
Concept drift detected in data: 7 - at index: 910
Concept drift detected in data: 9 - at index: 921
Concept drift detected in data: 8 - at index: 925

```
Concept drift detected in data: 9 - at index: 927
Concept drift detected in data: 7 - at index: 929
Warning detected in data: 9 - at index: 933
Concept drift detected in data: 9 - at index: 934
Concept drift detected in data: 9 - at index: 938
Concept drift detected in data: 7 - at index: 944
Concept drift detected in data: 8 - at index: 950
Concept drift detected in data: 9 - at index: 960
Warning detected in data: 7 - at index: 983
Warning detected in data: 7 - at index: 987
Concept drift detected in data: 10 - at index: 998
###############################################################
```

Drift detection using a moving weighted average-test (HDDM_W) is based on McDiarmid's bound [11]. McDiarmid's bound is a generalization of Hoeffding's bound. The HDDM_W involves the *W-test*, which involves weighted moving averages [8].

Listing 2-7 illustrates concept drift detection and warning zone detection using HDDM_W in scikit-multiflow. The experiment setup is the same as that of Listing 2-5. The only difference is that you are using the HDDM_W() method instead of HDDM_A(). The final output is shown in Listing 2-8.

Listing 2-7. Illustration for Concept Drift Detection Using HDDM_W in scikit-multiflow

```
###############################################################
# Import the relevant libraries
 import numpy as np
from skmultiflow.drift_detection.hddm_w import HDDM_W

# Initialize the HDDM_W object
HW = HDDM_W()
```

```
# set seed for reproducibility
np.random.seed(123)

# Simulate a data stream of size 1000 from a binomial distribution
# here, n= number of trials, p= probability of each trial
n, p = 10, 0.6
stream = np.random.binomial(n, p, 1000)

stream[:10]
# Output- array([5, 7, 7, 6, 5, 6, 3, 5, 6, 6])

# Data concept are changed from index 299 to 500
for j in range(299, 500):
    stream[j] = np.random.randint(5, high=9)

# Stream elements are added to DDM and checking whether drift
occurred
for j in range(1000):
    HW.add_element(stream[j])
    if HW.detected_change():
        print('Concept drift detected in data: ' +
        str(stream[j]) + ' - at index: ' + str(j))
    if HW.detected_warning_zone():
        print('Warning detected in data: ' + str(stream[j]) +
        ' - at index: ' + str(j))
##############################################################
```

Listing 2-8. Output for Concept Drift and Warning Zones Detected
Using HDDM_W

```
##############################################################
### Output:
Concept drift detected in data: 8 - at index: 13
Warning detected in data: 8 - at index: 30
Warning detected in data: 6 - at index: 31
```

Concept drift detected in data: 8 - at index: 41
Warning detected in data: 8 - at index: 65
Concept drift detected in data: 9 - at index: 74
Concept drift detected in data: 8 - at index: 96
Concept drift detected in data: 10 - at index: 118
Warning detected in data: 7 - at index: 122
Warning detected in data: 6 - at index: 123
Warning detected in data: 4 - at index: 124
Warning detected in data: 3 - at index: 125
Concept drift detected in data: 7 - at index: 126
Warning detected in data: 8 - at index: 135
Warning detected in data: 8 - at index: 140
Warning detected in data: 7 - at index: 141
Concept drift detected in data: 8 - at index: 147
Concept drift detected in data: 8 - at index: 150
Concept drift detected in data: 8 - at index: 155
Concept drift detected in data: 8 - at index: 163
Warning detected in data: 8 - at index: 171
Concept drift detected in data: 8 - at index: 175
Concept drift detected in data: 9 - at index: 194
Concept drift detected in data: 9 - at index: 196
Concept drift detected in data: 8 - at index: 199
Warning detected in data: 8 - at index: 206
Warning detected in data: 8 - at index: 209
Concept drift detected in data: 8 - at index: 225
Concept drift detected in data: 9 - at index: 235
Concept drift detected in data: 10 - at index: 240
Concept drift detected in data: 9 - at index: 258
Concept drift detected in data: 10 - at index: 272
Warning detected in data: 9 - at index: 279
Warning detected in data: 9 - at index: 281
Warning detected in data: 8 - at index: 282

```
Warning detected in data: 9 - at index: 284
Warning detected in data: 8 - at index: 294
Concept drift detected in data: 8 - at index: 298
Warning detected in data: 8 - at index: 353
Warning detected in data: 8 - at index: 355
Warning detected in data: 8 - at index: 378
Warning detected in data: 7 - at index: 379
Warning detected in data: 8 - at index: 380
Concept drift detected in data: 8 - at index: 381
Warning detected in data: 8 - at index: 464
Warning detected in data: 6 - at index: 465
Warning detected in data: 8 - at index: 469
Warning detected in data: 7 - at index: 470
Concept drift detected in data: 8 - at index: 471
Warning detected in data: 8 - at index: 486
Warning detected in data: 8 - at index: 496
Warning detected in data: 8 - at index: 497
Warning detected in data: 6 - at index: 498
Warning detected in data: 7 - at index: 499
Warning detected in data: 7 - at index: 500
Warning detected in data: 6 - at index: 501
Concept drift detected in data: 8 - at index: 523
Warning detected in data: 9 - at index: 530
Concept drift detected in data: 9 - at index: 547
Warning detected in data: 8 - at index: 554
Concept drift detected in data: 9 - at index: 561
Concept drift detected in data: 9 - at index: 566
Concept drift detected in data: 9 - at index: 586
Concept drift detected in data: 9 - at index: 598
Concept drift detected in data: 9 - at index: 644
Warning detected in data: 8 - at index: 670
Concept drift detected in data: 10 - at index: 671
```

```
Concept drift detected in data: 6 - at index: 673
Concept drift detected in data: 9 - at index: 703
Concept drift detected in data: 8 - at index: 705
Concept drift detected in data: 9 - at index: 716
Concept drift detected in data: 8 - at index: 723
Warning detected in data: 6 - at index: 751
Warning detected in data: 7 - at index: 752
Warning detected in data: 5 - at index: 753
Warning detected in data: 7 - at index: 754
Warning detected in data: 5 - at index: 755
Concept drift detected in data: 8 - at index: 756
Warning detected in data: 7 - at index: 771
Warning detected in data: 7 - at index: 772
Warning detected in data: 7 - at index: 776
Warning detected in data: 8 - at index: 777
Concept drift detected in data: 8 - at index: 788
Concept drift detected in data: 9 - at index: 802
Concept drift detected in data: 8 - at index: 814
Warning detected in data: 8 - at index: 826
Concept drift detected in data: 9 - at index: 864
Concept drift detected in data: 8 - at index: 867
Warning detected in data: 7 - at index: 873
Warning detected in data: 9 - at index: 874
Warning detected in data: 7 - at index: 875
Concept drift detected in data: 6 - at index: 876
Warning detected in data: 8 - at index: 905
Concept drift detected in data: 9 - at index: 934
Concept drift detected in data: 9 - at index: 960
Concept drift detected in data: 9 - at index: 975
Concept drift detected in data: 9 - at index: 982
Concept drift detected in data: 10 - at index: 998
###################################################################
```

An A-test is better at detecting abrupt or sudden concept drift; however, a W-test fares better in gradual concept drift.

Drift Detection Using the Page-Hinkley Method

The drift detection method based on the Page-Hinkley (PH) test [12] is a sequential analysis method that detects the sudden change in the mean of a Gaussian signal [13]. This method tracks the cumulated difference between the observed values and their mean up to the current moment [14]. Generally, this method sends an alarm for concept drift when the observed mean at some instant is greater than a given threshold λ. The value of λ is determined by the admissible false alarm rate. If you increase the value of λ, you may be able to reduce the false alarms, but you miss some instances of changes in the data distribution.

Listing 2-9 illustrates concept drift detection using the Page-Hinkley method. It simulates a data stream of size 1000 from a normal distribution and then changes the data concept from index 299 to 799 (i.e., replace values with integers between 5 and 8), as shown in Figure 2-2. However, unlike some of the methods discussed earlier, the Page-Hinkley method doesn't output warning zones. It only shows the instance of concept drift, as shown in Listing 2-10.

Listing 2-9. Illustration for Concept Drift Detection Using the Page-Hinkley Method in scikit-multiflow

```
###################################################################

# Import the relevant libraries
import numpy as np
from skmultiflow.drift_detection import PageHinkley

# Initialize the PageHinkley object
ph = PageHinkley()
```

```
# set seed for reproducibility
np.random.seed(123)
# Simulate a data stream of size 1000 from a normal
distribution
# with mean=0 and standard deviation=0.1
stream = np.random.normal(0, 0.1, 1000)

# Data concept are changed from index 299 to 799
for j in range(299, 800):
    stream[j] = np.random.randint(5, high=9)

# Adding stream elements to the PageHinkley drift detector and
verifying if drift occurred
for j in range(1000):
    ph.add_element(stream[j])
    if ph.detected_change():
        print('Concept drift detected in data: ' +
        str(stream[j]) + ' - at index: ' + str(j))
###############################################################
```

Listing 2-10. Output for Concept Drift Detected Using the Page-Hinkley Method

```
###############################################################
### Output:
Concept drift detected in data: 5.0 - at index: 306
###############################################################
```

	0
288	−0.132908
289	0.0278034
290	−0.107477
291	0.0668317
292	0.0955832
293	−0.0877614
294	−0.192372
295	0.0695787
296	0.18758
297	0.0415695
298	0.0160544
299	5
300	6
301	7
302	8
303	8
304	5
305	8
306	5
307	8
308	6

Figure 2-2. *The data stream with changed concept from index 299 to 799*

Summary

This chapter discussed the various algorithms for concept drift detection in a streaming data context. It also explored the relevant techniques and implemented them using the scikit-multiflow framework in Python.

The next chapter discusses various supervised learning techniques for streaming data and implements some of them using the scikit-multiflow framework.

References

[1] Shenghui Wang, Stefan Schlobach, and Michel Klein. "What Is Concept Drift and How to Measure It?" Knowledge Engineering and Management by the Masses. *Lecture Notes in Computer Science.* Springer, 2010.

[2] Michèle Basseville and Igor V. Nikiforov. *Detection of Abrupt Changes: Theory and Application.* Prentice-Hall, Inc., 1993.

[3] J. Lu, A. Liu, F. Dong, F. Gu, J. Gama, and G. Zhang. "Learning under Concept Drift: A Review." *IEEE Transactions on Knowledge & Data Engineering.* 2019.

[4] Albert Bifet and Ricard Gavalda. "Learning from time-changing data with adaptive windowing." *Proceedings of the 2007 SIAM international Conference on Data Mining.* Society for Industrial and Applied Mathematics. 2007. pp. 443-448.

[5]   Albert Bifet, Ricard Gavald, Geoff Holmes, and
      Bernhard Pfahringer. *Machine Learning for Data
      Streams: with Practical Examples in MOA*. MIT
      Press, 2018.

[6]   Joao Gama, Pedro Medas, Gladys Castillo, and Pedro
      Pereira Rodrigues. "Learning with drift detection."
      Advances in Artificial Intelligence: SBIA 2004. 17th
      Brazilian Symposium on Artificial Intelligence.
      September 29 – October 1, 2004. pp. 286–295.

[7]   Manuel Baena-Garcia, Jose Del Campo-Avila,
      Raúl Fidalgo, Albert Bifet, Ricard Gavaldà, Rafael
      Morales-Bueno. "Early Drift Detection Method."
      Fourth International Workshop on Knowledge
      Discovery from Data Streams. 2006.

[8]   Isvani Frías-Blanco, José del Campo-Ávila,
      Gonzalo Ramos-Jiménez, Rafael Morales-Bueno,
      Agustín Ortiz-Díaz, and Yailé Cabal. "Online and
      non-parametric drift detection methods based
      on Hoeffding's bounds." *IEEE Transactions on
      Knowledge and Data Engineering*. 2014.

[9]   Albert Bifet, Geoff Holmes, Richard Kirkby, and
      Bernhard Pfahringer. "MOA: Massive Online
      Analysis." *Journal of Machine Learning Research*.
      2010.

[10] W. Hoeffding. "Probabilities inequalities for sums
      of bounded random variables," Journal of the
      American Statistics Assoc. 1963. pp. 13– 30.

[11]   C. McDiarmid. "On the method of bounded differences." *Proc. Surv. Combinatorics*. 1989, pp. 148–188.

[12]   E. S. Page. "Continuous Inspection Schemes." *Biometrika*. 1954. pp. 100–115.

[13]   H. Mouss, D. Mouss, N. Mouss, and L. Sefouhi.. "Test of Page-Hinkley, an approach for fault detection in an agro-alimentary production system." Proceedings of the Asian Control Conference. 2004. pp. 815–818.

[14]   J. Gama. *Knowledge Discovery from Data Streams*. Chapman & Hall/CRC, 2010.

CHAPTER 3

Supervised Learning for Streaming Data

In the last chapter, you learned some concept drift detection algorithms. This chapter focuses on supervised learning algorithms (for classification tasks and regression tasks) in a streaming data context. In supervised learning, there is a target or a response variable and there are predictor variables. The goal of the learner/model is to understand the relationship between the target and the predictor variables. This can then be used to make predictions on the target variable for future observations of the predictor variables [1]. (I assume that you are already familiar with the supervised learning techniques used in a batch setting.)

Chapter 1 covered the challenges of streaming data and how the streaming data algorithms are different from batch learning algorithms. This chapter dives into some machine learning techniques for streaming data and their implementation using the scikit-multiflow framework.

In Chapter 1, you learned about data stream generators and how to create a data stream from a source file (CSV format). Those ideas are used in nearly all the working examples in this chapter.

The chapter starts with the evaluation methods used in a streaming data context. It then discusses supervised learning approaches—tree-based, lazy learning, Naïve Bayes, and more. Finally, the ensemble learning techniques in the streaming data context are explained.

© Sayan Putatunda 2021
S. Putatunda, *Practical Machine Learning for Streaming Data with Python*,
https://doi.org/10.1007/978-1-4842-6867-4_3

This chapter is restricted to binary classification methods and single target regression (i.e., multiclass classification and multitarget regression techniques are beyond the scope of this book). All the experiments in this chapter are performed on a system with a configuration of 1.6 GHz dual-core Intel Core i5 processor, 4 GB RAM, and macOS. However, the code should run on a similar or greater configuration system or operating system without any changes or modifications. This book uses the scikit-multiflow package version 0.5.3, which was released in June 2020.

Evaluation Methods

This section introduces model evaluation methods that are often used in evaluating the performance of supervised learning methods in a streaming data context. These methods are known as the *holdout evaluation method* and the *prequential evaluation method* (a.k.a., the *interleaved test-then-train method*).

In the holdout evaluation method, the performance evaluation happens after every batch of a certain number of examples or observations. The learner's or model's performance is evaluated on a test dataset formed by unseen examples. These unseen examples are used only for evaluation purposes and are not used to train the model. The same test dataset can be used in multiple evaluation runs. However, it is also possible to create dynamic test datasets. The holdout evaluation method can be applied using the EvaluateHoldout() function available in the scikit-multiflow framework's evaluation module.

The prequential evaluation method, or the interleaved test-then-train method, is specifically designed for evaluating models in a streaming data context. Each of the examples is analyzed according to its order of arrival. In this evaluation method, the incoming unseen example in the data stream is first used for testing (i.e., the model/learner makes predictions).

These unseen examples are also used to train the model before the next round of testing. The prequential evaluation method can be applied using the EvaluatePrequential() function in the scikit-multiflow framework's evaluation module.

There is another evaluation method available in the scikit-multiflow framework (the *prequential evaluation delayed method*); however, this book sticks to the holdout evaluation method and the prequential evaluation method, with more emphasis on the latter.

This chapter uses these evaluation methods to track the performance of various supervised learning algorithms discussed in the next few sections. You need to specify the metrics when using either the EvaluateHoldout() function or the EvaluatePrequential() function.

In this chapter, classification problems generally use metrics such as prediction accuracy and the f1 score (the harmonic mean of precision and recall). For regression problems, metrics such as mean squared error and mean absolute error are used.

Decision Trees for Streaming Data

This section discusses some of the tree-based techniques for classification and regression tasks in the streaming data context. It covers Hoeffding trees, Hoeffding Adaptive Tree classifiers, the Extremely Fast Decision Tree classifiers, Hoeffding tree regressors, and Hoeffding Adaptive Tree regressors, which are applied to different datasets using scikit-multiflow.

Hoeffding Tree Classifier

One of the challenges of streaming data is the memory requirements; it is impossible to store all the data. In a batch setting, a decision tree reuses instances or observations to compute the best split attributes. Thus, using decision tree methods such as CART and CHAID, which are generally used in batch data, aren't effective in a streaming data context. To solve this issue, Domingos and Hulten [2] proposed the Hoeffding tree, which is an incremental decision tree learner that uses new instances to build a tree, making it more suitable to a streaming data context. The Hoeffding tree is also known as the Very Fast Decision Tree (VFDT) algorithm.

The Hoeffding tree converges to a tree built by a batch learner with sufficiently large data. Thus, the Hoeffding tree uses the idea that a small sample is often enough for choosing an optimal splitting attribute [3]. This idea is supported by the statistical result known as the *Hoeffding bound*. The Hoeffding bound states (with a probability of 1–£) that after n independent observations of a real valued random variable, x, with range R, the true mean of x is at least $\mu - \epsilon$ where μ is the observed mean of the random variable x and $\epsilon = \sqrt{(R^2 \ln(1/\text{£})/2n)}$ [4].

The VFDT method is an implementation of Hoeffding trees. VFDT is a decision tree that uses the Hoeffding bound and accumulates sufficient statistics of new incoming observations in a data stream while converting a tree leaf to a tree node. However, streaming data can be very noisy, which affects the performance of VFDT (in terms of prediction accuracy). Also, with noisy incoming streaming data, VFDT faces tree-size explosion [5]. There are documented techniques to address these VFDT issues.

Listing 3-1 creates a synthetic dataset using the hyperplane generator discussed in Chapter 1. Figure 3-1 is a snapshot of the synthetic data stream.

The Hoeffding tree method is applied to the synthetic dataset using the scikit-multiflow framework, as shown in Listing 3-2. The prequential evaluation method (with prediction accuracy as a metric) is used in performance evaluation. Listing 3-3 shows the output of the prequential evaluation. The accuracy obtained is 92.07%.

Listing 3-1. Creating a Synthetic Dataset Using the Hyperplane Generator

```
###############################################################
# Import the relevant libraries
from skmultiflow.data import HyperplaneGenerator
import pandas as pd
import numpy as np

create = HyperplaneGenerator(random_state = 888, n_features=
10, noise_percentage = 0)
create.prepare_for_use()
X , Y = create.next_sample(10000)
data = pd.DataFrame(np.hstack((X, np.array([Y]).T)))

data.shape
# output- (10000, 11)
print(data.head())

# Store it in csv
data.to_csv('data_stream.csv', index=False)
###############################################################
```

Index	0	1	2	3	4	5	6	7	8	9	10
0	0.899470	0.248365	0.0301718	0.0724471	0.874184	0.55843	0.916047	0.63346	0.283253	0.365369	1
1	0.705171	0.648509	0.0409008	0.211732	0.00148992	0.130972	0.211825	0.0260949	0.446087	0.239105	0
2	0.0915874	0.977452	0.411501	0.458305	0.525909	0.294416	0.972116	0.181444	0.303406	0.174454	0
3	0.635272	0.496203	0.0141264	0.627222	0.634975	0.108141	0.829643	0.517752	0.570683	0.546333	1
4	0.450078	0.876507	0.537356	0.495684	0.418272	0.851006	0.386853	0.606895	0.217841	0.912944	1
5	0.856398	0.180293	0.948519	0.384163	0.251388	0.967406	0.77048	0.446852	0.198139	0.659823	1
6	0.578489	0.856236	0.357642	0.476238	0.545923	0.795083	0.144624	0.0180292	0.385322	0.906146	0
7	0.00179492	0.214237	0.28306	0.594816	0.42567	0.522077	0.780824	0.207883	0.768618	0.595955	0
8	0.175171	0.454754	0.461079	0.797812	0.0510449	0.424061	0.123532	0.916123	0.219467	0.0592571	0
9	0.563733	0.599512	0.166961	0.915744	0.226674	0.8908589	0.689444	0.391173	0.320342	0.722816	1
10	0.753456	0.809861	0.330064	0.197749	0.864116	0.962722	0.739514	0.674087	0.377465	0.794774	1
11	0.211648	0.824925	0.0736159	0.636996	0.303908	0.0696185	0.681359	0.56667	0.783518	0.0918758	0
12	0.625633	0.990047	0.393813	0.214151	0.693516	0.0356316	0.462469	0.537626	0.522928	0.464937	0
13	0.110446	0.92656	0.126959	0.846598	0.0312274	0.394362	0.582441	0.254377	0.298248	0.217418	0
14	0.25868	0.915166	0.0860515	0.723262	0.697547	0.957633	0.595555	0.194041	0.536536	0.567583	0

Figure 3-1. *Snapshot of the synthetic dataset*

Listing 3-2. Applying the Hoeffding Tree on the Synthetic Data Stream

```
################################################################
from skmultiflow.trees import HoeffdingTreeClassifier
from skmultiflow.evaluation import EvaluatePrequential
from skmultiflow.data.file_stream import FileStream
import pandas as pd
import numpy as np

# Load the synthetic data stream
dstream = FileStream('data_stream.csv')
dstream.prepare_for_use()

# Create the model instance
ht_class = HoeffdingTreeClassifier()

# perform prequential evaluation
evaluate1 = EvaluatePrequential(show_plot=False,
pretrain_size=400,
```

```
max_samples=10000,
metrics=['accuracy']
)
evaluate1.evaluate(stream=dstream, model=ht_class)
##############################################################
```

Listing 3-3. Prequential Evaluation Output of the Hoeffding Tree Classifier

```
##############################################################

Prequential Evaluation
Evaluating 1 target(s).
Pre-training on 400 sample(s).
Evaluating...
 ##----------------- [10%]
 ################### [100%] [3.95s]
Processed samples: 10000
Mean performance:
M0 - Accuracy     : 0.9207
##############################################################
```

Hoeffding Adaptive Tree Classifier

Chapter 2 introduced the adaptive windowing method for concept drift detection (ADWIN), in which a sliding window of dynamic size is used. The window size is not fixed; it is computed online based on the rate of change observed from the data in the window. The ADWIN algorithm increases the window size if it doesn't see any change in the data stream and reduces the window size when there's a change detected in the data stream.

The Hoeffding Adaptive Tree uses the ADWIN method and is a modified version of the Hoeffding tree. First, you need to understand the Concept Adapting Very Fast Decision Tree (CVFDT) method [6]. The CVFDT method is more appropriate for data streams with inherent concept drift. It updates statistics at inner nodes and leaves. The core idea is that whenever a change is detected at a subtree, it grows a candidate subtree, and eventually, either the current subtree or the candidate subtree is dropped.

The Hoeffding Adaptive Tree generates an alternate decision tree at nodes where the splitting test is no longer appropriate [6]. The old tree is replaced by a new one, which is more accurate. The decisions made on the leaf are based on a recent examples/observations window. Memory needs to be allocated for an example window; however, there's no rigorous performance guarantee with CVFDT.

The Hoeffding Adaptive Tree is a slight modification of the CVFDT, which uses ADWIN. The purpose of ADWIN is to monitor the error of each subtree and the alternate trees. The Hoeffding Adaptive Tree method uses the ADWIN estimates to make decisions on the leaf and on growing new trees or alternate trees. This method doesn't need to allocate memory for an example window; it is almost as accurate as of the CVFDT and sometimes even performs better. Also, a rigorous performance guarantee is possible.

Listing 3-4 simulates a synthetic data stream with concept drift using the Agrawal generator. The Hoeffding Adaptive Tree classifier is applied to the synthetic data stream and evaluates the model's performance using both a prequential evaluation and a Holdout evaluation.

Listing 3-5 shows the output of the prequential evaluation approach. The accuracy obtained is 99.72%, whereas in the holdout approach, the accuracy obtained is 99.09% (as shown in Listing 3-6).

Listing 3-4. Applying the Hoeffding Adaptive Tree Classifier on a Synthetic Data Stream

```
###############################################################
# Import the relevant libraries
from skmultiflow.trees import HoeffdingAdaptiveTreeClassifier
from skmultiflow.data import ConceptDriftStream
from skmultiflow.evaluation import EvaluatePrequential
from skmultiflow.evaluation import EvaluateHoldout

# Simulate a sample data stream
ds = ConceptDriftStream(random_state=777, position=30000)
ds
# Output:
#ConceptDriftStream(alpha=0.0,
#          drift_stream=AGRAWALGenerator(balance_classes=False,
#                                        classification_function=2,
#                                        perturbation=0.0,
#                                        random_state=112),
#          position=30000, random_state=777,
#          stream=AGRAWALGenerator(balance_classes=False,
#                                  classification_function=0,
#
#                  perturbation=0.0, random_state=112),
#          width=1000)

# Instantiate the model object
model_hat = HoeffdingAdaptiveTreeClassifier()

# Prequential evaluation
eval1 = EvaluatePrequential(pretrain_size=400, max_samples=300000,
batch_size=1,n_wait=100, max_time=2000,show_plot=False,
metrics=['accuracy'])
```

```
eval1.evaluate(stream=ds, model=model_hat)

# Holdout evaluation
eval2 = EvaluateHoldout(max_samples=30000,
                        max_time=2000,
                        show_plot=False,
                        metrics=['accuracy'],
                        dynamic_test_set=True)

eval2.evaluate(stream=ds, model=model_hat)
################################################################
```

Listing 3-5. Prequential Evaluation Output of the Hoeffding Adaptive Tree Classifier

```
################################################################
Prequential Evaluation
Evaluating 1 target(s).
Pre-training on 400 sample(s).
Evaluating...
 #################### [100%] [342.55s]
Processed samples: 300000
Mean performance:
M0 - Accuracy      : 0.9972
################################################################
```

Listing 3-6. Holdout Evaluation Output of the Hoeffding Adaptive Tree Classifier

```
################################################################
Holdout Evaluation
Evaluating 1 target(s).
Evaluating...
```

```
######-------------- [30%] [6.87s]Separating 5000 holdout
samples.
###############---- [80%] [15.71s]Separating 5000 holdout
samples.
Processed samples: 30000
Mean performance:
M0 - Accuracy     : 0.9909
##############################################################
```

Extremely Fast Decision Tree Classifier

The Extremely Fast Decision Tree (EFDT) (an instantiation of the Hoeffding Anytime Tree) is an incremental decision tree. The Hoeffding Anytime Tree operates almost like a Hoeffding tree, but the difference lies in the way they split at a node. The Hoeffding tree delays the split at a node until it identifies the best split and doesn't revisit the decision. The Hoeffding Anytime Tree splits at a node as soon as it seems to be a useful split and does revisit the decision in the availability of a better split. The Hoeffding Anytime Tree is not computationally as efficient as the Hoeffding tree, but it is statistically more efficient [7].

Listing 3-7 uses the same synthetic data stream used in Listing 3-4. It applies the EFDT classifier to the synthetic dataset and evaluates the model's performance using both the prequential evaluation and the holdout evaluation. Listing 3-8 shows the output of the prequential evaluation approach. The accuracy obtained is 96.07%, whereas in the holdout approach, the accuracy obtained is 96.75% (see Listing 3-9).

Listing 3-7. Applying the Extremely Fast Decision Tree Classifier on a Synthetic Data Stream

```
##############################################################
# Import the relevant libraries
from skmultiflow.trees import ExtremelyFastDecisionTreeClassifier
from skmultiflow.data import ConceptDriftStream
from skmultiflow.evaluation import EvaluatePrequential
from skmultiflow.evaluation import EvaluateHoldout

# Simulate a sample data stream
ds = ConceptDriftStream(random_state=777, position=30000)
ds
# Output:
#ConceptDriftStream(alpha=0.0,
#        drift_stream=AGRAWALGenerator(balance_classes=False,
#                                      classification_function=2,
#                                      perturbation=0.0,
#                                      random_state=112),
#        position=30000, random_state=777,
#        stream=AGRAWALGenerator(balance_classes=False,
#                                classification_function=0,
#                                perturbation=0.0, random_state=112),
#        width=1000)

# Instantiate the model object
model_hat = ExtremelyFastDecisionTreeClassifier()

# Prequential evaluation
eval1 = EvaluatePrequential(pretrain_size=400, max_samples=300000,
batch_size=1,
                n_wait=100, max_time=2000,
                        show_plot=False, metrics=['accuracy'])
```

```
eval1.evaluate(stream=ds, model=model_hat)

# Holdout evaluation
eval2 = EvaluateHoldout(max_samples=30000,
                        max_time=2000,
                        show_plot=False,
                        metrics=['accuracy'],
                        dynamic_test_set=True)

eval2.evaluate(stream=ds, model=model_hat)
###############################################################
```

Listing 3-8. Prequential Evaluation Output of the Extremely Fast
Decision Tree Classifier

```
###############################################################

Prequential Evaluation
Evaluating 1 target(s).
Pre-training on 400 sample(s).
Evaluating...
 #####--------------- [25%] [1820.77s]
Time limit reached (2000.00s). Evaluation stopped.
Processed samples: 80949
Mean performance:
M0 - Accuracy      : 0.9607
###############################################################
```

Listing 3-9. Holdout Evaluation Output of the Extremely Fast
Decision Tree Classifier

```
################################################################

Holdout Evaluation
Evaluating 1 target(s).
Evaluating...
 ######-------------- [30%] [267.24s]Separating 5000 holdout
 samples.
 ################---- [80%] [505.29s]Separating 5000 holdout
 samples.
Processed samples: 30000
Mean performance:
M0 - Accuracy     : 0.9675
################################################################
```

Hoeffding Tree Regressor

The Hoeffding tree regressor is similar to the Hoeffding tree classifier. It
uses the Hoeffding bound for making split decisions. The Hoeffding tree
regressor uses reduction in variance in the target space for deciding on the
split candidates. It makes predictions by fitting a linear perceptron model
or computing the sample average.

Listing 3-10 creates a synthetic data stream for regression using the
RegressionGenerator() function in the scikit-multiflow framework. The
data stream (dstream) has 800 observations, nine features, and one target
variable. The target variable is continuous because this is a regression
context. You can see a particular sample in the dstream object using the
next_sample() function. The Hoeffding tree regressor is applied to the data
stream, and a prequential evaluation is applied.

The metrics used in the regression context are the *mean squared error* (MSE) and the *mean absolute error* (MAE). Listing 3-11 shows the output of the prequential evaluation. The MSE is 6665.2407, and the MAE is 62.974944.

Listing 3-10. Applying the Hoeffding Tree Regressor on a Synthetic Data Stream

```
###############################################################
# Import the relevant libraries
from skmultiflow.trees import HoeffdingTreeRegressor
from skmultiflow.data import RegressionGenerator
import pandas as pd
from skmultiflow.evaluation import EvaluatePrequential

# Setup a data stream
dstream = RegressionGenerator(n_features=9, n_samples=800,
n_targets=1, random_state=456)

dstream

dstream.next_sample()
#(array([[ 0.72465838, -1.92979924, -0.02607907,  2.35603757,
          -0.37461529,
#         -0.38480019,  0.06603468, -2.1436878 ,  0.49182531]]),
# array([61.302191]))

# Instantiate the Hoeffding Tree Regressor object
htr = HoeffdingTreeRegressor()
```

```
# Prequential evaluation
eval1 = EvaluatePrequential(pretrain_size=400, max_samples=800,
batch_size=1, n_wait=100, max_time=2000, show_plot=False,
metrics=['mean_square_error', 'mean_absolute_error'])

eval1.evaluate(stream=dstream, model=htr)
################################################################
```

Listing 3-11. Prequential Evaluation Output of the Hoeffding Tree Regressor

```
################################################################
Prequential Evaluation
Evaluating 1 target(s).
Pre-training on 400 sample(s).
Evaluating...
 ################### [100%] [0.69s]
Processed samples: 799
Mean performance:
M0 - MSE         : 6665.2407
M0 - MAE         : 62.974944
################################################################
```

Hoeffding Adaptive Tree Regressor

The Hoeffding Adaptive Tree can be used in a regression context. The Hoeffding Adaptive Tree regressor uses the ADWIN method and can be considered a modified version of the Hoeffding tree regressor.

Listing 3-12 uses the same synthetic dataset used in Listing 3-10. The Hoeffding Adaptive Tree regressor is applied to this dataset, and a prequential evaluation is performed. The output of the prequential evaluation is shown in Listing 3-13. The MSE is 6222.9041, and the MAE is 61.320054.

Listing 3-12. Applying the Hoeffding Adaptive Tree Regressor to a Synthetic Data Stream

```
###############################################################
# Import the relevant libraries
from skmultiflow.trees import HoeffdingAdaptiveTreeRegressor
from skmultiflow.data import RegressionGenerator
import pandas as pd
from skmultiflow.evaluation import EvaluatePrequential

# Setup a data stream
dstream = RegressionGenerator(n_features=9, n_samples=800,
n_targets=1, random_state=456)

dstream

dstream.next_sample()
#(array([[ 0.72465838, -1.92979924, -0.02607907,  2.35603757,
         -0.37461529,
#         -0.38480019,  0.06603468, -2.1436878 ,  0.49182531]]),
# array([61.302191]))

# Instantiate the Hoeffding Tree Regressor object
model_hatr = HoeffdingAdaptiveTreeRegressor()

# Prequential evaluation
eval1 = EvaluatePrequential(pretrain_size=400, max_samples=800,
batch_size=1, n_wait=100, max_time=2000, show_plot=False,
metrics=['mean_square_error', 'mean_absolute_error'])

eval1.evaluate(stream=dstream, model=model_hatr)
###############################################################
```

Listing 3-13. Prequential Evaluation Output of the Hoeffding Adaptive Tree Regressor

```
###############################################################

Prequential Evaluation
Evaluating 1 target(s).
Pre-training on 400 sample(s).
Evaluating...
#################### [100%] [1.23s]
Processed samples: 799
Mean performance:
M0 - MSE        : 6222.9041
M0 - MAE        : 61.320054
###############################################################
```

This section discussed some of the prominent tree-based techniques for classification and regression in a streaming data context. The next section focuses on some of the lazy learning techniques.

Lazy Learning Methods for Streaming Data

Let's talk a bit about the k-nearest neighbor (KNN) method for classification and regression in a batch context. KNN is a non-parametric method that keeps all the observations in memory and classifies a new incoming observation based on a similarity metric/distance function [9]. Euclidean distance, Mahalanobis distance, and the Manhattan distance are three of the commonly used distance functions.

In a new observation, the KNN algorithm selects the nearest k neighbors in the dataset based on the distance function and then combines the values of the k nearest neighbors for prediction (i.e., in a regression context, it takes the average, and in a classification context, it performs majority voting).

scikit-multiflow has an implementation of the KNN method for classification in streaming data context (KNNClassifier()). The modus-operandi of this technique is the same as that of KNN, along with a window on the training samples. The user can fix the maximum size of the window. The KNNClassifier() method treats all features in a data stream as continuous and is not recommended for datasets with a mix of categorical and continuous features.

Listing 3-14 applies the KNN classifier (where K = 10) on a synthetic dataset, and prequential evaluation is applied. The same synthetic dataset generated using HyperplaneGenerator() in Listing 3-1 is exported to a CSV file (data_stream.csv). A snapshot of the dataset is shown in Figure 3-1. The output of the prequential evaluation is shown in Listing 3-15. The accuracy obtained is 87.99%.

Listing 3-14. Applying the KNN Classifier to a Synthetic Data Stream

```
###############################################################
# Import the relevant libraries
from skmultiflow.lazy import KNNClassifier
from skmultiflow.evaluation import EvaluatePrequential
from skmultiflow.data.file_stream import FileStream
import pandas as pd
import numpy as np

dstream = FileStream('data_stream.csv')
dstream.prepare_for_use()

# Instantiate the KNN Classifier method
knn_class = KNNClassifier(n_neighbors=10, max_window_size=1000)
```

```
# Prequential Evaluation
evaluate1 = EvaluatePrequential(show_plot=False,
pretrain_size=1000,
max_samples=10000,
metrics=['accuracy']
)
# Run the evaluation
evaluate1.evaluate(stream=dstream, model=knn_class)
#############################################################
```

Listing 3-15. Prequential Evaluation Output of the KNN Classifier

```
#############################################################

Prequential Evaluation
Evaluating 1 target(s).
Pre-training on 1000 sample(s).
Evaluating...
 ##----------------- [10%]
 #################### [100%] [7.31s]
Processed samples: 10000
Mean performance:
M0 - Accuracy    : 0.8799
#############################################################
```

An improvement over the KNN classifier is the KNN ADWIN classifier (i.e., KNNADWINClassifier()), which handles the concept drift in a data stream more effectively than a KNN classifier. This technique uses the adaptive windowing method for concept drift detection (ADWIN).

Listing 3-16 creates a synthetic dataset using the SEA generator available in the scikit-multiflow framework. The SEA generator creates a data stream with concept drift and generates three numerical features and a binary target variable. SEA is the acronym for *streaming ensemble algorithm* [10]. Listing 3-16 shows an output of five observations of the simulated data stream. The KNN ADWIN classifier (K = 10) is applied to the data stream, and prequential evaluation is applied. The output of the prequential evaluation is shown in Listing 3-17. The accuracy obtained is 65.93%.

Listing 3-16. Applying the KNN ADWIN Classifier to a Synthetic Data Stream

```
################################################################
# Import the relevant libraries
from skmultiflow.lazy import KNNADWINClassifier
from skmultiflow.evaluation import EvaluatePrequential
from skmultiflow.data.sea_generator import SEAGenerator

# Simulate the data stream
dstream = SEAGenerator(classification_function = 2, balance_
classes = True, noise_percentage = 0.3, random_state = 333)

#Retrieve five samples
dstream.next_sample(5)
# Output:
#(array([[3.68721825, 0.48303666, 1.04530188],
#        [2.45403315, 8.73489354, 0.51611639],
#        [2.38740114, 2.03699194, 1.74533621],
#        [9.41738118, 4.66915281, 9.59978205],
#        [1.05404748, 0.42265956, 2.44130999]]),
array([1, 0, 0, 1, 1]))

# Instantiate the KNN ADWIN classifier method
adwin_knn_class = KNNADWINClassifier(n_neighbors=10,
max_window_size=1000)
```

```
# Prequential Evaluation
evaluate1 = EvaluatePrequential(show_plot=False,
pretrain_size=1000,
max_samples=10000,
metrics=['accuracy'])

# Run the evaluation
evaluate1.evaluate(stream=dstream, model=adwin_knn_class) #####
####################################################################
```

Listing 3-17. Prequential Evaluation Output of the KNN ADWIN Classifier

```
####################################################################

Prequential Evaluation
Evaluating 1 target(s).
Pre-training on 1000 sample(s).
Evaluating...
 #################### [100%] [11.22s]
Processed samples: 10000
Mean performance:
M0 - Accuracy     : 0.6593
####################################################################
```

Now, let's look at another type of KNN-based classifier: the Self Adjusting Memory (SAM) model for the KNN algorithm [11]. This method is known as SAM-KNN for the rest of this chapter. The SAM-KNN technique effectively handles the heterogeneous concept drift and doesn't require optimization of metaparameters. This technique derives inspiration from the dual-store model in human memory research, where knowledge is partitioned into short-term memory (STM) and long-term memory (LTM) [12].

In SAM-KNN, memory is partitioned into STM and LTM. STM contains data related to the recent concepts in a dynamic sliding window. LTM maintains knowledge of past information. There is a cleaning process to keep the size of the STM under control. Moreover, there is a transfer of filtered knowledge from STM to LTM. LTM ensures that none of is compressed stored information contradicts that of STM. The combined memory (CM) is the union of STM and LTM.

The idea is to combine/ensemble dedicated models (distance weighted KNN) for the latest concepts and all the past concepts to maximize the prediction accuracy.

Listing 3-18 applies the SAM-KNN classifier and performs prequential evaluation on the same data stream used in Listing 3-16. In Listing 3-19, the accuracy obtained is 64.66%.

Listing 3-18. Applying the SAM-KNN Classifier to a Synthetic Data Stream

```
###############################################################
# Import the relevant libraries
from skmultiflow.lazy import SAMKNNClassifier
from skmultiflow.evaluation import EvaluatePrequential
from skmultiflow.data.sea_generator import SEAGenerator

# Simulate the data stream
dstream = SEAGenerator(classification_function = 2, balance_
classes = True, noise_percentage = 0.3, random_state = 333)

#Retrieve five samples
dstream.next_sample(5)

# Instantiate the SAM-KNN classifier method
sam_knn_class = SAMKNNClassifier(n_neighbors=10,
weighting='distance', stm_size_option='maxACCApprox',
max_window_size=1000,use_ltm=True)
```

```
# Prequential Evaluation
evaluate1 = EvaluatePrequential(show_plot=False,
pretrain_size=1000,
max_samples=10000,
metrics=['accuracy']
)
# Run the evaluation
evaluate1.evaluate(stream=dstream, model=sam_knn_class)
################################################################
```

Listing 3-19. Prequential Evaluation Output of the SAM-KNN Classifier

```
################################################################

Prequential Evaluation
Evaluating 1 target(s).
Pre-training on 1000 sample(s).
Evaluating...
 #################### [100%] [8.53s]
Processed samples: 10000
Mean performance:
M0 - Accuracy     : 0.6466
################################################################
```

scikit-multiflow has an implementation of the KNN method for regression in a streaming data context (KNNRegressor()). Just like the KNN classifier, a window on the training samples is used. When compared to that of the KNN classifier, however, it takes the average of the k nearest neighbors for prediction.

Listing 3-20 applies the KNN regressor to a synthetic dataset. It uses the same synthetic dataset used in Listing 3-10. A prequential evaluation is applied, and the output is shown in Listing 3-21. The MSE is 4195.3502, and the MAE is 50.132250.

Listing 3-20. Applying the KNN Regressor to a Synthetic Data Stream

```
##############################################################
# Import the relevant libraries
from skmultiflow.lazy import KNNRegressor
from skmultiflow.data import RegressionGenerator
import pandas as pd
from skmultiflow.evaluation import EvaluatePrequential

# Setup a data stream
dstream = RegressionGenerator(n_features=9, n_samples=800,
n_targets=1, random_state=456)

dstream

dstream.next_sample()
#(array([[ 0.72465838, -1.92979924, -0.02607907,  2.35603757,
          -0.37461529,
#          -0.38480019,  0.06603468, -2.1436878 ,  0.49182531]]),
# array([61.302191]))

# Instantiate the KNN Regressor object
knn_reg = KNNRegressor()

# Prequential evaluation
eval1 = EvaluatePrequential(pretrain_size=400, max_samples=800,
batch_size=1, n_wait=100, max_time=2000, show_plot=False,
metrics=['mean_square_error', 'mean_absolute_error'])

eval1.evaluate(stream=dstream, model=knn_reg)
##############################################################
```

Listing 3-21. Prequential Evaluation Output of the KNN Regressor

```
################################################################

Prequential Evaluation
Evaluating 1 target(s).
Pre-training on 400 sample(s).
Evaluating...
 ################### [100%] [0.34s]
Processed samples: 799
Mean performance:
M0 - MSE          : 4195.3502
M0 - MAE          : 50.132250
################################################################
```

The next section focuses on ensemble learning for streaming data.

Ensemble Learning for Streaming Data

Ensemble learning combines multiple models to improve the prediction accuracy for out-of-sample data. Typically (though not guaranteed), an ensemble learner performs better than the stand-alone base methods. Ensemble learning is a very popular technique in machine learning for both regression and classification purposes. Some of the well-known ensemble learning techniques in a batch data context are random forests, bagging, and boosting. This section focuses on some of the ensemble learning techniques for streaming data implemented in the scikit-multiflow framework.

Adaptive Random Forests

A *random forest* is a very popular and widely used technique for machine learning on batch (non-streaming) data. In a random forest, multiple decision trees (built using resampled data and with a random subset of features) are combined to improve a model's performance in terms of prediction accuracy. However, random forests can't handle concept drift and require multiple passes over the dataset. It is not suitable for analysis in a streaming data context.

An *adaptive random forest* (ARF) is an adaptation of the random forest technique in a streaming data context [13]. The ARF method uses online bagging to approximate resampling with replacement. However, instead of using a ($\lambda = 1$) Poisson distribution in online bagging, ARF uses a ($\lambda = 6$) Poisson distribution. (An overview of the online bagging method is presented later in this chapter.)

The base model in ARF (i.e., random forest tree training (RFTreeTrain)) is based on the Hoeffding tree algorithm, but it has a few differences. One difference is that the RFTreeTrain doesn't allow early tree pruning. Another difference is that whenever a new node is created, a random subset of features is created, and splits are limited to only these features. The ARF uses drift detectors in base trees that cause resets whenever there is a drift. ARF allows training background trees, which replace active trees if a drift is detected.

The ARF technique can be used in classification tasks and regression tasks. Listing 3-22 illustrates applying the ARF classifier on a synthetic data stream (the same data stream used in Listing 3-18). The hyperparameter for the ARF classifier is the number of estimators/trees is 100. The maximum number of features is the square root of the total number of features in the data stream. A prequential evaluation is applied. Listing 3-23 shows that the prediction accuracy is 70.18%.

Listing 3-22. Applying the ARF Classifier to a Synthetic Data Stream

```
###############################################################
# Import the relevant libraries
from skmultiflow.meta import AdaptiveRandomForestClassifier
from skmultiflow.evaluation import EvaluatePrequential
from skmultiflow.data.sea_generator import SEAGenerator
```

```
# Simulate the data stream
dstream = SEAGenerator(classification_function = 2, balance_
classes = True, noise_percentage = 0.3, random_state = 333)

# Instatntiate the KNN ADWIN classifier method
ARF_class = AdaptiveRandomForestClassifier(n_estimators=100,
max_features="sqrt", random_state=333)

# Prequential Evaluation
evaluate1 = EvaluatePrequential(show_plot=False,
pretrain_size=1000,
max_samples=10000,
metrics=['accuracy']
)
# Run the evaluation
evaluate1.evaluate(stream=dstream, model=ARF_class)
#############################################################
```

Listing 3-23. Prequential Evaluation Output of the ARF Classifier

```
#############################################################

Prequential Evaluation
Evaluating 1 target(s).
Pre-training on 1000 sample(s).
Evaluating...
 ################### [100%] [1526.63s]
Processed samples: 10000
Mean performance:
M0 - Accuracy     : 0.7018
#############################################################
```

Listing 3-24 applies the ARF in a regression context on the same synthetic data stream generated in Listing 3-20. The number of estimators is 100, and the maximum number of features is the square root of the total number of features. A prequential evaluation is applied, and the output is shown in Listing 3-25. The MSE is 3661.6862, and the MAE is 46.579340.

Listing 3-24. Applying the ARF Regressor on a Synthetic Data Stream

```
###############################################################
# Import the relevant libraries
from skmultiflow.meta import AdaptiveRandomForestRegressor
from skmultiflow.data import RegressionGenerator
from skmultiflow.evaluation import EvaluatePrequential

# Setup a data stream
dstream = RegressionGenerator(n_features=9, n_samples=800,
n_targets=1, random_state=456)

# Instantiate the ARF Regressor object
ARF_reg = AdaptiveRandomForestRegressor(n_estimators=100,
max_features="sqrt", random_state=333)

# Prequential evaluation
eval1 = EvaluatePrequential(pretrain_size=400, max_samples=800,
batch_size=1, n_wait=100, max_time=2000, show_plot=False,
metrics=['mean_square_error', 'mean_absolute_error'])

eval1.evaluate(stream=dstream, model=ARF_reg)
###############################################################
```

Listing 3-25. Prequential Evaluation Output of the ARF Regressor

```
###############################################################

Prequential Evaluation
Evaluating 1 target(s).
Pre-training on 400 sample(s).
Evaluating...
```

```
#################### [100%] [77.83s]
Processed samples: 800
Mean performance:
M0 - MSE          : 3661.6862
M0 - MAE          : 46.579340
###############################################################
```

Online Bagging

Bagging is a widely used ensemble learning technique in batch (non-streaming) learning. Here, X base models are trained on bootstrap sample sets of size N. The bootstrapped sample sets are created using sampling with replacement from the original training dataset.

The training set of each of the base models contains the original training examples K times, where $P(K = k)$ follows a binomial distribution. And when N tends to be infinity, the distribution of K tends to be a Poisson(1) distribution (i.e., $K \sim \exp(-1)/k!$) [15]. Oza and Russell [15] used this idea to propose the online learning version of the bagging algorithm, which is suitable in a streaming data context.

In online bagging, as a new training example arrives, the algorithm chooses the example $K \sim$ Poisson(1) times for each base model and updates the base model accordingly. New instances are classified by performing majority voting of the X base models. This step is the same for both batch and online bagging methods. Online bagging is considered a good approximation to batch bagging [15].

In scikit-multiflow, the online bagging classification technique can be applied using the OzaBaggingClassifier() method. Listing 3-26 applies the Oza bagging classifier to the synthetic dataset used in Listing 3-16. The KNN ADWIN classifier model is the base model. Listing 3-26 built six base models. The hyperparameters of the KNN ADWIN classifier are the same as that used in Listing 3-16 (i.e., number of neighbors is 10 and maximum window size is 1000). A prequential evaluation is applied, and the output is shown in Listing 3-27. The accuracy obtained is 66.21%.

Listing 3-26. Applying the Oza Bagging Classifier to a Synthetic Data Stream

```
###############################################################
# Import the relevant libraries
from skmultiflow.meta import OzaBaggingClassifier
from skmultiflow.lazy import KNNADWINClassifier
from skmultiflow.evaluation import EvaluatePrequential
from skmultiflow.data.sea_generator import SEAGenerator

# Simulate the data stream
dstream = SEAGenerator(classification_function = 2, balance_
classes = True, noise_percentage = 0.3, random_state = 333)

# Instantiate the Oza Bagging classifier method with KNN ADWIN
classifier as the base model
oza_class = OzaBaggingClassifier(base_
estimator=KNNADWINClassifier(n_neighbors=10, max_window_
size=1000), n_estimators=4, random_state = 333)

# Prequential Evaluation
evaluate1 = EvaluatePrequential(show_plot=False,
pretrain_size=1000,
max_samples=10000,
metrics=['accuracy']
)
# Run the evaluation
evaluate1.evaluate(stream=dstream, model=oza_class)
###############################################################
```

Listing 3-27. Prequential Evaluation Output of the Oza Bagging
Classifier

```
################################################################

Prequential Evaluation
Evaluating 1 target(s).
Pre-training on 1000 sample(s).
Evaluating...
 ################### [100%] [61.53s]
Processed samples: 10000
Mean performance:
M0 - Accuracy    : 0.6621
################################################################
```

Now, let's talk about the online bagging method known as the
leveraging bagging method, proposed by Bifet et al. [16]. Here, bagging is
leveraged by increasing resampling and using output detection codes. In
the Oza Bagging technique, a sampling with replacement uses the ($\lambda = 1$)
Poisson distribution. But in the leveraging bagging method, the value of
λ is larger. For example, instead of using the ($\lambda = 1$) Poisson distribution,
it uses the ($\lambda = 6$) Poisson distribution. The output detection code helps
correct errors. Each class is assigned a binary string of length L, and then
an ensemble of L binary classifiers is built. This leads to better performance
in multiclass classification tasks. Also, the leveraging bagging algorithm
uses the ADWIN windowing technique for handling inherent concept drift
in a data stream.

In scikit-multiflow, the leveraging bagging classification technique can
be applied using the LeveragingBaggingClassifier() method. Listing 3-28
replicates the same experiment performed in Listing 3-26 (i.e., the same
data stream and hyperparameters for the six base models (KNN ADWIN
classifiers)). However, the OzaBaggingClassifier() method is replaced with

the LeveragingBaggingClassifier() method. A prequential evaluation is performed, and the output is shown in Listing 3-29. The accuracy obtained is 58.68%.

Listing 3-28. Applying the Leveraging Bagging Classifier to a Synthetic Data Stream

```
###############################################################
# Import the relevant libraries
from skmultiflow.meta import LeveragingBaggingClassifier
from skmultiflow.lazy import KNNADWINClassifier
from skmultiflow.evaluation import EvaluatePrequential
from skmultiflow.data.sea_generator import SEAGenerator

# Simulate the data stream
dstream = SEAGenerator(classification_function = 2, balance_
classes = True, noise_percentage = 0.3, random_state = 333)

# Instantiate the Leveraging Bagging classifier method with KNN
ADWIN classifier as the base model
leverage_class = LeveragingBaggingClassifier(base_
estimator=KNNADWINClassifier(n_neighbors=10, max_window_
size=1000), n_estimators=6, random_state = 333)

# Prequential Evaluation
evaluate1 = EvaluatePrequential(show_plot=False,
pretrain_size=1000,
max_samples=10000,
metrics=['accuracy']
)
# Run the evaluation
evaluate1.evaluate(stream=dstream, model=leverage_class)
###############################################################
```

Listing 3-29. Prequential Evaluation Output of the Leveraging
Bagging Classifier

```
################################################################

Prequential Evaluation
Evaluating 1 target(s).
Pre-training on 1000 sample(s).
Evaluating...
 #################### [100%] [310.46s]
Processed samples: 10000
Mean performance:
M0 - Accuracy    : 0.5868
################################################################
```

Online Boosting

Another well-known ensemble learning technique is *boosting*. Its basic
principle is to sequentially combine a set of "weak learners" to create
a "strong learner." The AdaBoost technique is one of the first boosting
techniques for batch data [17]. (AdaBoost is short for *adaptive boosting*.)
A weak classifier (i.e., a decision tree) is prepared on training data using
weighted samples, and then the misclassification error rate is calculated
for the trained model. The misclassified examples by the current classifier
are given more weight in the training dataset for the next classifier. The
subsequent learners focus more on the difficult examples, which goes
on until there is no more significant improvement in the algorithm's
performance.

Wang and Pineau [18] proposed the *online boosting classifier*, which
is the online version of the AdaBoost algorithm. In an online learning
context, each arriving example is trained K times, drawn from a binomial
distribution. As in a data stream, the number of examples is large (tends
to infinity), so the binomial distribution tends to a Poisson(λ) distribution

where the value of λ is calculated by keeping track of the weight of misclassified and correctly classified examples. The online boosting algorithm uses the ADWIN windowing technique for handling inherent concept drift in a data stream.

The online AdaBoost algorithm can be applied using the OnlineBoostingClassifier() method implemented in the scikit-multiflow framework. Listing 3-30 applies the online boosting classifier on the same data stream used in Listing 3-28. A prequential evaluation is performed, and the output is shown in Listing 3-31. The accuracy obtained is 58%.

Listing 3-30. Applying the Online Boosting Classifier to a Synthetic Data Stream

```
###############################################################
# Import the relevant libraries
from skmultiflow.meta import OnlineBoostingClassifier
from skmultiflow.evaluation import EvaluatePrequential
from skmultiflow.data.sea_generator import SEAGenerator

# Simulate the data stream
dstream = SEAGenerator(classification_function = 2, balance_
classes = True, noise_percentage = 0.3, random_state = 333)

# Instantiate the Online Boosting Classifier method
boost_class = OnlineBoostingClassifier(random_state = 333)

# Prequential Evaluation
evaluate1 = EvaluatePrequential(show_plot=False,
pretrain_size=1000,
max_samples=10000,
metrics=['accuracy']
```

```
)
# Run the evaluation
evaluate1.evaluate(stream=dstream, model=boost_class)
###################################################################
```

Listing 3-31. Prequential Evaluation Output of the Online Boosting
Classifier

```
###################################################################
Prequential Evaluation
Evaluating 1 target(s).
Pre-training on 1000 sample(s).
Evaluating...
 ################### [100%] [669.82s]
Processed samples: 10000
Mean performance:
M0 - Accuracy     : 0.5800
###################################################################
```

Most of the important supervised learning algorithms in the streaming
data context have been covered. The next section briefly covers some of
the data stream preprocessing methods.

Data Stream Preprocessing

So far, this chapter has covered algorithms for both regression and
classification problems on data streams. All the experiments used clean
datasets (i.e., data preprocessing wasn't necessary). However, you can't
expect clean data streams in the real world, so you need to do some
basic data stream preprocessing. The following are some of the data
preprocessing methods available in the scikit-multiflow framework to
apply to a data stream.

- Missing Value Cleaner: The MissingValueCleaner() method can replace missing values in a data stream using strategies such as replacing missing values with zero, mean, median, or mode.

- One-hot encoding: The OneHotToCategorical() method performs one-hot encoding on categorical variables in a data stream.

- Windowed Minmax Scaler: The WindowedMinmaxScaler(window_size=200) method is for feature transformation on a data stream. Each feature is scaled to a given range (between 0 and 1). The window size is user input to compute the min and the max values.

- Windowed Standard Scaler: The WindowedStandardScaler(window_size=200) method standardizes the features. The window size is user input to compute the mean and the standard deviation values.

Summary

This chapter covered most of the important online supervised learning techniques for data stream mining in the scikit-multiflow framework. However, you should explore two methods that I didn't discuss in this chapter: the Naïve Bayes classifier and the Perceptron classifier available in the scikit-multiflow's Bayes module and the scikit-multiflow framework's neural network module.

The next chapter discusses some of the online/incremental learning algorithms for unsupervised learning of streaming data. It also discusses some other tools available for streaming data analysis.

References

[1]   Gareth James, Daniela Witten, Trevor Hastie, and
      Robert Tibshirani. An *Introduction to Statistical
      Learning with Applications in R*. Springer Texts in
      Statistics, 2013.

[2]   P. Domingos and G. Hulten. *Mining High-Speed
      Data Streams*. ACM Press, 2000.

[3]   Albert Bifet and Ricard Gavaldà. *Adaptive Learning
      from Evolving Data Streams*. Springer, 2009.

[4]   G. Hulten, L. Spencer, and P. Domingos. *Mining
      Time-Changing Data Streams*. ACM Press, 2001.

[5]   H. Yang and S. Fong. "Optimized very fast decision
      tree with balanced classification accuracy
      and compact tree size." The 3rd International
      Conference on Data Mining and Intelligent
      Information Technology Applications. IEEE, 2011.

[6]   Geoff Hulten, Laurie Spencer, and Pedro Domingos.
      "Mining Time-Changing Data Streams." Proceedings
      of the 7th ACM SIGKDD International conference
      on Knowledge Discovery and Data Mining. ACM
      Press, 2001.

[7]   C. Manapragada, G. Webb, and M. Salehi.
      "Extremely Fast Decision Tree." Proceedings of the
      24th ACM SIGKDD International Conference on
      Knowledge Discovery & Data Mining. https://doi.
      org/10.1145/3219819.3220005

[8]   Elena Ikonomovska, João Gama, Raquel Sebastião,
      and Dejan Gjorgjevik. "Regression Trees from Data

Streams with Drift Detection." *Discovery Science.* Springer, 2009.

[9]   Kilian Q. Weinberger and Lawrence K. Saul. "Distance Metric Learning for Large Margin Nearest Neighbor Classification." *Journal of Machine Learning Research*, 2009.

[10]   W. Nick Street and YongSeog Kim. "A streaming ensemble algorithm (SEA) for large-scale classification." Proceedings of the seventh ACM SIGKDD international conference on Knowledge discovery and data mining. ACM, 2001.

[11]   Viktor Losing, Barbara Hammer, and Heiko Wersing. "KNN Classifier with Self Adjusting Memory for Heterogeneous Concept Drift." IEEE 16th International Conference on Data Mining (ICDM). IEEE, 2016.

[12]   R. C. Atkinson and R. M. Shiffrin, "Human Memory: A Proposed System and Its Control Processes." *Psychology of Learning and Motivation*, 1968. pp. 89–195.

[13]   Heitor Murilo Gomes, Albert Bifet, Jesse Read, Jean Paul Barddal, Fabricio Enembreck, Bernhard Pfharinger, Geoff Holmes, and Talel Abdessalem. "Adaptive random forests for evolving data stream classification." *Machine Learning*. Springer, 2017.

[14]   Heitor Murilo Gomes, Jean Paul Barddal, Luis Eduardo Boiko Ferreira, and Albert Bifet. "Adaptive random forests for data stream regression." ESANN, 2018.

[15]   Nikunj C. Oza and Stuart Russell. "Online Bagging and Boosting." *Artificial Intelligence and Statistics.* January 2001, pp. 105–112.

[16]   A. Bifet, G. Holmes, and B. Pfahringer, "Leveraging Bagging for Evolving Data Streams." Joint European Conference on Machine Learning and Knowledge Discovery in Databases, 2010. pp. 135–150.

[17]   Yoav Freund and Robert E. Schapire. "A decision-theoretic generalization of online learning and an application to boosting." *Computational Learning Theory.* Springer, 1995.

[18]   B. Wang and J. Pineau, "Online Bagging and Boosting for Imbalanced Data Streams." *IEEE Transactions on Knowledge and Data Engineering.* December 2016. pp. 3353-3366.

CHAPTER 4

Unsupervised Learning and Other Tools for Data Stream Mining

In Chapter 3, you learned supervised machine learning techniques for both regression and classification problems in a streaming data context. This chapter starts with unsupervised learning for streaming data and then overviews some of the other software environments available for data stream mining.

Unsupervised Learning for Streaming Data

In the previous chapters, you learned about the various methods for concept drift detection and supervised machine learning for streaming data. Let's now look at unsupervised learning by first focusing on clustering and then moving on to anomaly detection.

© Sayan Putatunda 2021
S. Putatunda, *Practical Machine Learning for Streaming Data with Python*,
https://doi.org/10.1007/978-1-4842-6867-4_4

Clustering

k-means clustering is a popular clustering algorithm for batch data. It partitions the data into distinct and non-overlapping k clusters. The objective is to have a very low *within-cluster variation* [1]. The algorithm randomly assigns a number from 1 to k to the observations. This creates the initial set of clusters. The cluster centroids are computed for each of the k clusters. Observations based on the nearness (i.e., a distance metric, such as Euclidean distance) to a cluster centroid are reassigned. These steps are reiterated. The iteration stops when there are no further changes in the cluster assignments. (Please note that the user provides the value of k.)

The *mini-batch k-means clustering method* is an incremental learning version of the k-means clustering algorithm that you can directly apply to a data stream. It is a variant of the k-means algorithm that works on a mini-batch of the data instead of the entire dataset. It is suitable for streaming data since an entire dataset is never available at any point in time. Mini-batches are the subsets of the input data stream. They are randomly sampled for each training iteration.

The mini-batch k-means clustering algorithm chooses a mini-batch comprising randomly sampled examples. A generic k-means algorithm is then executed, in which the nearest centroid is assigned. In the next step, the assigned centroids are updated for each sample in the mini-batch. These steps are performed until either convergence or a specific number of iterations is reached.

Also, instead of using random sampling to choose the initial set of cluster centers, the k-means++ algorithm [3] can be used as a randomized seeding technique to choose the initial cluster centers' first values, which achieves convergence faster.

The mini-batch k-means clustering algorithm is available in the scikit-learn (sklearn) library. Go to `https://scikit-learn.org/stable/modules/generated/sklearn.cluster.MiniBatchKMeans.html` for more information.

Listing 4-1 applies the mini-batch k-means clustering on the Iris dataset. The Iris dataset is publicly available in the sklearn library. The dataset contains four features regarding sepal and petal length (in cm) and width (in cm). Listing 4-1 applies **MiniBatchKMeans()** and specifies the number of clusters as 3 and the mini-batch size as 50. Listing 4-1 trained the mini-batch k-means clustering model on the Iris dataset and shown the cluster centers in Figure 4-1.

Listing 4-1. Mini-Batch k-means Clustering on the Iris Dataset

```
################################################################
# Import the relevant libraries
from sklearn import datasets
from sklearn.preprocessing import StandardScaler
from sklearn.cluster import MiniBatchKMeans

# Load data
iris = datasets.load_iris()
X = iris.data
#Data snapshot

X[0:6,]
#Output
#array([[5.1, 3.5, 1.4, 0.2],
#       [4.9, 3. , 1.4, 0.2],
#       [4.7, 3.2, 1.3, 0.2],
#       [4.6, 3.1, 1.5, 0.2],
#       [5. , 3.6, 1.4, 0.2],
#       [5.4, 3.9, 1.7, 0.4]])

#feature names
iris.feature_names
#Output:
#['sepal length (cm)',
```

```
# 'sepal width (cm)',
# 'petal length (cm)',
# 'petal width (cm)']

# Create k-mean object
clust_kmeans = MiniBatchKMeans(n_clusters=3, batch_size=50,
random_state=333)

# Train model
model = clust_kmeans.fit(X)

# Figure out the cluster centers
model.cluster_centers_
###############################################################
        array([[5.97476415, 2.77051887, 4.48396226, 1.47122642],
               [4.97687688, 3.37987988, 1.46696697, 0.24384384],
               [6.87489712, 3.1       , 5.75226337, 2.1037037 ]])
```

Figure 4-1. *Cluster venters of the mini-batch k-means clustering model*

An alternative to mini-batch k-means clustering is available at scikit-learn. This method is known as *Birch clustering* [7]. It is an online learning algorithm that creates a tree data structure where the cluster centroids are read off the leaf. I suggest that you explore this technique as an exercise. For more information on Birch clustering in scikit-learn, read the official documentation at https://scikit-learn.org/stable/modules/generated/sklearn.cluster.Birch.html.

A *streaming k-means clustering algorithm* is available in Apache Spark's machine learning library, MLlib. This streaming k-means clustering algorithm dynamically estimates the clusters and updates the clusters when new data arrive. This algorithm contains a parameter called a *half-life* that controls the "decay" of the estimates. A decay factor determines how much weight needs to be given to past data. The streaming k-means clustering algorithm uses a generalization of the mini-batch k-means update rule.

Other incremental clustering algorithms are not covered in this book.

Some of the earliest online clustering algorithms for streaming data are implemented in MOA software. One well-researched approach for data stream clustering is the microclustering algorithm, which has two phases. The first phase is online and summarizes a data stream in local models (microclusters). The second phase generates a global cluster model from the micro-clusters. CluStream and DenStream are two of the microcluster-based online clustering algorithms available in MOA.

Apache Spark, MOA, and a few other data stream mining tools are discussed next.

Anomaly Detection

Anomaly detection (or *outlier detection*) is the science of identifying events or examples that differ from the norm [4] (known as *anomalies*). Machine learning literature is replete with techniques for anomaly detection, especially in a batch learning context. Some of the unique characteristics associated with these anomalies are that they rarely occur in a dataset and their features are different from normal examples. Some of the practical applications of anomaly detection are intrusion detection, fraud detection, surveillance analytics, medical applications, and data leakage prevention.

This chapter focuses on the *streaming half-space trees* (SHST) technique for anomaly detection in a streaming data context. The SHST algorithm is an incremental one-class anomaly detector for streaming data [5]. One of the most useful features of this algorithm is that it can be trained with only genuine or normal instances and works well even when anomalies are extremely rare. This method is a random ensemble of half-space trees. A half-space tree (HS-tree) of depth d is defined as a full binary tree having $2^{(d+1)} - 1$ nodes, where all leaves are at the same depth (i.e., d).

Streaming half-space trees can create the tree structure from the data space dimensions alone and don't need actual training examples. So, this method doesn't need to perform any model restructuring to adapt to the streaming data. The model can be deployed before the arrival of streaming data. Moreover, the SHST algorithm has constant memory requirements and constant amortized time complexity.

Listing 4-2 generates a data stream with anomalies using AnomalySineGenerator, available in the scikit-multiflow framework. It generates a data stream with 500 observations, out of which 100 are anomalies. A snapshot of the generated data stream is shown in Figure 4-2. The rightmost column indicates whether an example is anomalous (i.e., if its 1, it is an anomaly; if it's 0, it is normal). It then applies the SHST algorithm on this data stream using the HalfSpaceTrees() method available in the scikit-multiflow framework. It trains the model and performs a prediction, as shown in Listing 4-2.

Listing 4-2. Anomaly Detection Using Streaming Half-Space Trees

```
##################################################################
# Import the relevant libraries
import pandas as pd
from skmultiflow.anomaly_detection import HalfSpaceTrees
from skmultiflow.data import AnomalySineGenerator
```

```
# Generate a data stream with anomalies
dstream = AnomalySineGenerator(n_samples=500, n_anomalies=100,
random_state=333)

# Instantiate the Half-Space Trees estimator
HS_trees = HalfSpaceTrees(random_state=333)

#prep the data stream
dstream.prepare_for_use()
X, Y = dstream.next_sample(500)
data = pd.DataFrame(np.hstack((X, np.array([Y]).T)))
data.head()

# Incrementally fit the model
HS_trees_model= HS_trees.partial_fit(X, Y)

# Predict the classes of the passed observations
pred = HS_trees_model.predict(X)

# Estimate the probability of a sample being anomalous or
  genuine
prob = HS_trees_model.predict_proba(X)
##################################################################
```

Index	0	1	2
0	0.858592	1.41399	1
1	0.409751	1.48899	0
2	0.242154	0.202679	0
3	0.61268	0.843635	0
4	1.10394	−0.0498022	0
5	0.770067	−0.400269	0
6	0.633393	0.78983	0
7	0.662943	−0.562824	0
8	0.750774	1.17144	1
9	0.168572	−0.613221	0
10	0.572482	1.07164	1
11	0.728675	−0.73326	0
12	0.0793211	−0.270515	0

Figure 4-2. *Snapshot of the data stream with anomalies*

Other Tools and Technologies for Data Stream Mining

This section provides a high-level introduction to eight open source software tools for data stream mining: MOA, Apache Spark, Apache Flink, Apache Storm, Apache Kafka, Faust, Creme, and River. There are other tools available (both commercial and open source) for data stream mining, but they are outside the scope of this book.

Massive Online Analysis (MOA)

Massive Online Analysis (MOA) is an open source framework for mining data streams developed in Java. MOA is a GUI-based tool. It contains a collection of streaming data algorithms for supervised (classification, regression, etc.) and unsupervised learning (clustering, etc.) along with model evaluation tools [6]. Most of the supervised and unsupervised learning discussed in this book are available in MOA.

Since MOA is GUI-based and primarily developed in Java, it may not appeal to Python developers/users. I recommend that you explore this tool. It can be downloaded from https://moa.cms.waikato.ac.nz.

Apache Spark

Apache Spark is a software environment for performing high-speed processing and analytics on both batch and streaming data. It provides support for multiple languages such as Java, Python, and R. That's why Spark (or pyspark) is a good alternative option for a Python user. Spark can access diverse data sources. It can run in the cloud or on Mesos, Hadoop, and Kubernetes. Spark's stack of libraries includes Spark SQL, Spark Streaming, MLlib (for machine learning), and GraphX (for graph analytics). To download the latest version and to access the documentation, please visit https://spark.apache.org.

Spark Streaming is an extension of the core Spark API. It is a powerful tool for processing data streams at scale. It contains many high-level functions, such as map, reduce, join, and window for processing high-velocity data streams. Spark Streaming supports MLlib operations (i.e., streaming k-means clustering, streaming linear regression (a linear regression model with stochastic gradient descent fitted for each batch of the data), and more). The official Spark Streaming guide is at https://spark.apache.org/docs/latest/streaming-programming-guide.html.

Apache Flink

Apache Flink is a unified batch and streaming data analytics engine. It provides highly flexible streaming windows for a continuous streaming model. It can perform in-memory computations at scale. Flink is integrated with many other open source data processing ecosystems, such as YARN and Mesos. Flink is very useful for event-driven applications such as anomaly detection, rule-based alerts, fraud detection, and business process monitoring. Please visit `https://flink.apache.org` to learn more about Apache Flink.

Apache Storm

Apache Storm is an open source distributed stream-processing computation framework. It is primarily developed in Clojure but works with multiple languages, such as Java and Python. Apache Storm is a very powerful tool for performing real-time processing on streaming data. It is a stream processing engine without batch support. It is a true real-time processing framework, taking in a stream as an entire event instead of series of small batches. It is scalable and works on parallel calculations that run across a cluster of machines.

A Storm cluster is similar to what a Hadoop cluster is for batch data. To perform real-time computations in Storm, you create a *topology*, which is a directed acyclic graph of computation. A topology is made of *streams* (i.e., an unbounded pipeline of tuples) and *spouts* (i.e., source of the data stream). A Storm cluster comprises two nodes: master nodes and worker nodes. The master node assigns tasks to machines and monitors their performance. A worker node assign tasks to other worker nodes.

To download the latest version of Apache Storm and access the official documentation, please visit `https://storm.apache.org/index.html`.

Apache Kafka

Apache Kafka is an open source distributed event streaming platform. What is *event streaming*? It is the phenomenon of capturing real-time data streams for various sources (databases, sensors, etc.) and then storing and processing them to generate insights. Some of the real-world use-cases for event streaming are processing financial transactions in stock exchanges, analyzing IoT device data, self-driving applications, web analytics, log aggregation, and real-time patient health monitoring. The official documentation is available at `https://kafka.apache.org/documentation/`.

Apache Kafka introduced a powerful but lightweight data stream processing library known as Kafka Streams. Developers can build Java and Scala applications at scale using Kafka Streams. Applications built using it can be deployed to VMs, containers, or the cloud. Documentation on Kafka Streams is available at `https://kafka.apache.org/documentation/streams/`.

Faust

Faust is a library for building streaming applications in Python. It is useful for event streaming and is similar to Kafka Streams. But Faust lets you use Python syntax and other popular libraries in Python, such as NumPy, TensorFlow, and Django, for processing data streams. The official documentation is available at `https://faust.readthedocs.io/en/latest/`. You can also visit its GitHub page at `https://github.com/robinhood/faust`.

Creme

Creme is an open source library for online machine learning in Python. It is a powerful tool used to learn from streaming data. It shares many similarities with the scikit-multiflow framework. Creme can be installed from PyPI using the following command.

```
>>> pip install creme
```

Please visit Creme's GitHub page of to explore this tool further- https://github.com/MaxHalford/creme.

River

River [8] is a new library for online machine learning of streaming data using Python. It is the result of the merger of Creme and scikit-multiflow. At the time of writing this chapter (November 2020), the developers of both packages (Creme and scikit-multiflow) announced the merger and attested the common shared vision and goals. Figures 4-3 and 4-4 show the merger announcement on the scikit-multiflow and Creme GitHub pages. Both of these screenshots/images were taken on 22nd Novemeber, 2020.

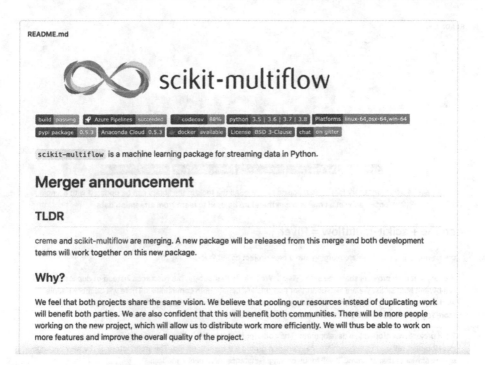

Figure 4-3. *Screenshot of the merger announcement on scikit-multiflow's GitHub page*

README.md

Creme

`build` `passing` `coverage` `84%` `release` `v0.6.1` `downloads` `112497` `License` `BSD 3-Clause`

`creme` is a Python library for online machine learning. All the tools in the library can be updated with a single observation at a time, and can therefore be used to **learn from streaming data**.

creme + scikit-multiflow = River

creme and scikit-multiflow are merging into a new project called River.

We feel that both projects share the same vision. We believe that pooling our resources instead of duplicating work will benefit both sides. We are also confident that this will benefit both communities. There will be more people working on the new project, which will allow us to distribute work more efficiently. We will thus be able to work on more features and improve the overall quality of the project.

Both projects will stop active development. The code for both projects will remain publicly available, although development will only focus on minor maintenance during a transition period. The architecture of the new package is very similar to that of creme. It will focus on single-instance incremental models.

We encourage users to use River instead of creme. We understand that this transition will require an extra effort in the short term from current users. However, we believe that the result will be better for everyone in the long run.

You will still be able to install and use `creme` as well as `scikit-multiflow`. Both projects will remain on PyPI, conda-forge and GitHub.

Figure 4-4. *Screenshot of the merger announcement in Creme's GitHub page*

The new package resulting from the merger is known as River. Creme and scikit-multiflow remain on PyPI, GitHub, and conda-forge. The River package is available in GitHub at `https://github.com/online-ml/river`. The API reference document is available at `https://riverml.xyz/latest/api/overview/`.

At the time of writing this chapter, the River package was too new to discuss anything further. Many features still need to be developed. I believe that River has the potential to be a one-stop solution in Python for online machine learning with streaming data. Maybe I can focus on this package in the next edition of this book.

110

Conclusion and the Path Forward

The purpose of this book is to introduce online/incremental machine learning with streaming data using Python. The real-time or near real-time analytics is the *next big thing* (especially with the advent of IoT devices, wearables, sensors, etc.). Many organizations and research institutes have are focusing on this.

Chapter 1 explained how machine learning with streaming data is different. Many machine learning methods that most of us use daily in a batch context aren't effective for data stream mining.

This book introduced the various online/incremental learning algorithms for concept drift detection, supervised learning, and unsupervised learning with streaming data. Given the demand of these skills, I strongly believe that these techniques should be in the repertoire of any data scientist, researcher, machine learning engineers, analytics professional, or data science aspirants.

The book primarily focused on a Python-based library for data stream processing (i.e., the scikit-multiflow framework). This makes sense given the fact that Python is the most widely used language by data scientists. I discussed quite a few other tools in this chapter, and there are plenty of books and other resources available (especially on Spark, Kafka, and MOA). However, few books deal with machine learning with streaming data. This book is a step in this direction.

Scikit-multiflow is a very powerful tool for online/incremental machine learning with streaming data. You used it to implement various supervised and unsupervised learning techniques in a streaming data context in this book.

I recommend that you start with scikit-multiflow to learn and explore the online/incremental machine learning techniques discussed in this book. Again, the package has merged with the Creme library, and the new package is known as River. I suggest that you transition to the River library once you are comfortable with the nitty-gritty of the scikit-multiflow framework. Please note that most of the techniques you learned in the scikit-multiflow framework should be integrated into the River package.

Again, at the time of writing this chapter, the River package is too new to discuss in more detail. But I may cover this package or any other package that comes up as a suitable alternative to scikit-multiflow in the next edition of this book.

The path forward is to learn the nuances of various online/incremental machine learning algorithms using the scikit-multiflow framework on various data streams and then move on to explore other Python-based tools for data stream mining, such as River, Spark, and Faust.

Happy learning! ☺

References

[1]   Leonard Kaufman and Peter Rousseeuw. *Finding Groups in Data: An Introduction to Cluster Analysis.* Wiley, 1990.

[2]   D. Sculley. "Web-Scale K-Means Clustering." WWW '10: Proceedings of the 19th international conference on World Wide Web. April 2010.

[3]   Bahman Bahmani, Benjamin Moseley, Andrea Vattani, Ravi Kumar, and Sergei Vassilvitskii. "Scalable K-Means++." Proceedings of the VLDB Endowment. March 2012.

[4]   Markus Goldstein and Seiichi Uchida. "A Comparative
       Evaluation of Unsupervised Anomaly Detection
       Algorithms for Multivariate Data." Wayne State
       University. April 19, 2016. `https://journals.`
       `plos.org/plosone/article?id=10.1371/journal.`
       `pone.0152173.`

[5]   S. C. Tan, K. M. Ting, and T. F. Liu, "Fast anomaly
       detection for streaming data." IJCAI Proceedings,
       International Joint Conference on Artificial
       Intelligence, 2011.

[6]   A. Bifet, G. Holmes, R. Kirkby, and B. Pfahringer,
       "MOA: Massive Online Analysis." *Journal of Machine
       Learning Research*, 2010.

[7]   Tian Zhang, Raghu Ramakrishnan, and Maron
       Livny. "BIRCH: An efficient data clustering
       method for large databases." `www.cs.sfu.ca/`
       `CourseCentral/459/han/papers/zhang96.pdf.`

[8]   Jacob Montiel, Max Halford, Saulo Martiello
       Mastelini, Geoffrey Bolmier, Raphael Sourty, Robin
       Vaysse, Adil Zouitine, Heitor Murilo Gomes, Jesse
       Read, Talel Abdessalem, and Albert Bifet, "River:
       machine learning for streaming data in Python.",
       arXiv, 2020.

Index

Printed in the United States
by Baker & Taylor Publisher Services